Studies in Choice and Welfare

M. Salles (Editor-in-Chief and Series Editor)

P. K. Pattaniak (Series Editor)

K. Suzumura (Series Editor)

Studies in Choice and Welfare

D. Austen-Smith and *J. Duggan*
Social Choice and Strategic Decisions
XVI, 319 pages, 2005. ISBN 3-540-22053-4

W. Kuklys
Amartya Sen's Capability Approach
XVIII, 116 pages, 2005. ISBN 3-540-26198-2

Bruno Simeone · Friedrich Pukelsheim
(Editors)

Mathematics and Democracy

Recent Advances in Voting Systems
and Collective Choice

With 33 Figures and 50 Tables

 Springer

Prof. Dr. Bruno Simeone
Department of Statistics
University of Rome »La Sapienza«
Piazzale Aldo Moro 5
00185 Rome, Italy
bruno.simeone@uniroma1.it

Prof. Dr. Friedrich Pukelsheim
Institute of Mathematics
University of Augsburg
86135 Augsburg, Germany
Pukelsheim@Math.Uni-Augsburg.De

ISSN 1614-0311
ISBN-10   3-540-35603-7 Springer Berlin Heidelberg New York
ISBN-13   978-3-540-35603-5 Springer Berlin Heidelberg New York

Cataloging-in-Publication Data
Library of Congress Control Number: 2006932296

Springer is a part of Springer Science+Business Media

springeronline.com

© Springer Berlin · Heidelberg 2006

Hardcover-Design: Erich Kirchner, Heidelberg

SPIN 11782865   43/3100-5 4 3 2 1 0 – Printed on acid-free paper

Preface

Voting Systems and Collective Choice are subjects at the crossroad of social and exact sciences. Social sciences look at the development and evolution of electoral systems and other rules of collective choice, in the context of the changing needs and political patterns of society. Exact sciences are concerned instead with the formal study of voting mechanisms and other preference aggregation procedures, under axioms which are meant to reflect such universal principles as equity, representation, stability, and consistency.

Mathematics and Democracy presents a collection of research papers focussing on quantitative aspects of electoral system theory, such as game-theoretic, decision-theoretic, statistical, probabilistic, combinatorial, geometric, and optimization-based approaches. Electronic voting protocols and other security issues have also recently been devoted a great deal of attention. Quantitative analyses provide a powerful tool to detect inconsistencies or poor performance in actual systems. The topics covered are at the forefront of the research in the field:

- Proportional methods

- Biproportional apportionment

- Approval voting

- Gerrymandering

- Metric approaches to Social Choice

- Impossibility theorems

Applications to concrete cases such as the procedures used to elect the EU Parliament, the US Congress, and various national and regional assemblies are discussed, as well as issues related to committee voting.

Mathematicians, economists, statisticians, computer scientists, engineers, and quantitatively oriented political and social scientists will find this book not only of the highest quality, but also eclectic, discussion-oriented, and mind-provoking. Altogether, *Mathematics and Democracy* offers an appraisal from leading specialists of "what's new" in the emerging multidisciplinary science of voting systems and social choice, with a broad view on real-life applications.

The present book grew out of the *International Workshop on Mathematics and Democracy: Voting Systems and Collective Choice*, which took place 18–23 September 2005 at the Ettore Majorana Centre in the charming medieval town of Erice, Sicily, under our joint scientific directorship. The Workshop's aim was to bring together different viewpoints on the subject, and to stress the role of mathematics towards a deeper understanding, a rational assessment, and a sound design of voting procedures. The invited speakers, representing many countries, are prominent scholars from their disciplines. Many of the papers included here were first presented at the Workshop, while some were adjoined later on. All papers were refereed; we would like to thank the referees for their indispensible contributions.

The idea of the Workshop goes back as far as to the year 1999. As Bruno Simeone recalls: "In the Spring 1999 I had the chance to organize a Mini-symposium on Electoral Systems at DIMACS. There I met people like Steve Brams and Don Saari, and the idea of having a workshop in Erice came up. That Summer, I was enjoying my vacations in Santa Marinella, a seaside town north of Rome, when I got a phone call from Professor Antonino Zichichi, the well-known physicist who heads the Erice Majorana Centre: *How about giving a talk in Erice on Mathematics and Democracy in front of a few Nobel Laureates?* The title of the Workshop—and of this book—was born at that very moment. Needless to say, my vacations dissolved and I gave that talk. In Erice, I proposed to Professor Zichichi the idea of a Workshop on Voting Systems and he instantly endorsed it. He envisaged in the Workshop an excellent opportunity to encourage a multi-disciplinary debate among the most qualified international experts of electoral systems and to promote exchange between the academic world and the wider society in order to disseminate scientific findings which are of collective interest, according to the well-established tradition of the Centre. However, due to the casual intertwining of many events, some years elapsed before the project actually materialized. The turning point was the Oberwolfach Workshop on electoral systems organized by Michel Balinski, Steven Brams, and Friedrich Pukelsheim. There I had the opportunity to ask many distinguished participants whether they would be interested in a second Workshop to take place in Erice. Their massive favourable reaction convinced me to go ahead and do it."

Drawing up the conclusions of the Workshop and relying on input from several invited speakers, Professor Michel Balinski wrote down a "Declaration" for the proper choice of an electoral system. This document, now better known as The *Erice Decalogue* and already translated in other languages, was unanimously undersigned by the participants. We are particularly pleased to include in this volume (pages xi–xii) such an authoritative document providing sound guidelines for electoral reform planning and advocating quantitative methods for electoral system assessment.

Neither the Workshop nor the book would have been possible without the intervention of many persons and institutions. We are most grateful to Professor Antonino Zichichi, President of the *Ettore Majorana Foundation and Centre for Scientific Culture*, for his enthusiastic support and encouragement. We acknowledge the indispensable financial and logistic support of the Centre and its staff, headed by Dr. Fiorella Ruggiu.

We are grateful to Professor Renato Guarini, Rector of La Sapienza University; Professor Gabriella Salinetti, Dean of the Faculty of Statistical Sciences; and to Professor Paolo Dell'Olmo, Head of the Statistics Department, for their patronage and financial help. Additional funding was made available by the Italian Ministry of University and Research, and by the Sicily Region.

We are much indebted to Professor Maurice Salles and Dr. Martina Bihn of Springer-Verlag, for their prompt endorsement of our proposal to publish this book in the Springer series *Studies in Choice and Welfare*.

Our warmest thanks to the other two formidable members of the Organizing Committee, Federica Ricca and Aline Pennisi, for their precious, ubiquitous, and clever assistance in the Workshop organization and logistics. They were efficiently backed by Isabella Lari and Andrea Scozzari, to whom we extend our thanks. Sebastian Maier and Federica Ricca have done a fantastic and supersonic job in editing the L^AT_EX format of the articles for the book.

BRUNO SIMEONE AND FRIEDRICH PUKELSHEIM

Rome and Augsburg, July 2006

Contents

Preface   v

The Erice Decalogue   xi

Power Indices Taking into Account Agents' Preferences   1
Fuad Aleskerov

The Sunfish Against the Octopus: Opposing Compactness to Gerrymandering   19
Nicola Apollonio, Ronald I. Becker, Isabella Lari, Federica Ricca, Bruno Simeone

Apportionment: Uni- and Bi-Dimensional   43
Michel Balinski

Minimum Total Deviation Apportionments   55
Paul H. Edelman

Comparison of Electoral Systems: Simulative and Game Theoretic Approaches   65
Vito Fragnelli, Guido Ortona

How to Elect a Representative Committee Using Approval Balloting   83
D. Marc Kilgour, Steven J. Brams, M. Remzi Sanver

On Some Distance Aspects in Social Choice Theory   97
Christian Klamler

Algorithms for Biproportional Apportionment   105
Sebastian Maier

Distance from Consensus: A Theme and Variations   117
Tommi Meskanen, Hannu Nurmi

A Strategic Problem in Approval Voting   133
Jack H. Nagel

The Italian Bug: A Flawed Procedure for Bi-Proportional Seat Allocation   151
Aline Pennisi

Current Issues of Apportionment Methods   167
Friedrich Pukelsheim

A Gentle Majority Clause for the Apportionment of Committee Seats   177
Friedrich Pukelsheim, Sebastian Maier

Allotment According to Preferential Vote: Ecuador's Elections   189
Victoriano Ramírez

Degressively Proportional Methods for the Allotment of the European
Parliament Seats Amongst the EU Member States   205
Victoriano Ramírez, Antonio Palomares, Maria L. Márquez

Hidden Mathematical Structures of Voting   221
Donald G. Saari

A Comparison of Electoral Formulae for the Faroese Parliament   235
Petur Zachariassen, Martin Zachariassen

List of Talks   253

List of Participants   255

The Erice Decalogue

The objective of the scientific research presented and discussed at the International Workshop on *Mathematics and Democracy: Voting Systems and Collective Choice* is the development and understanding of fair electoral systems. Pursued by scientists coming from different disciplines – mathematics, political science, economics, the law, computer science, ... – there is a common set of principles that the participants share.

We believe that a fair electoral system for electing a Parliament should:

1 Ensure transparency and simplicity. Voting systems whose properties are simple to understand by the electorate should be preferred to complex ones, and they should respect a nation's historical and legal context.

2 Guarantee accuracy. The act of voting – with paper ballots, optical scanners, electronic or other devices – should be able to assure to voters that their votes were accurately counted.

3 Promote competitiveness and avoid partisan bias. The system should favor no political group over another. In particular, it should render (almost) impossible the election of a majority in the Parliament with a minority of the voters.

4 Make every vote count. A system should never discourage a citizen from voting; it must encourage participation.

5 Make the Parliament a "mirror" of the electorate representing the divergent "popular wills," yet capable of governing (through, for example, the emergence of a majority).

6 Minimize the incentives to vote strategically. The system should encourage voters to express sincerely their true preferences.

7 Eliminate partisan political control by assigning the legal and administrative responsibility for elections to an independent commission.

A system using electoral districts should:

8 Encourage geographical compactness of the districts and respect natural geographical features and barriers.

9 Respect existing political subdivisions and communities of interest, and make every effort to avoid confusion among districts defined for different elections (local, regional and national).

10 Guarantee redistricting on a regular basis to account for demographic changes (but never in response to partisan appetites); at same time, it should recognize the limited precision and transitory nature of census data.

Theory is necessary to understand the properties and consequences of choosing one or another electoral system. A "science" of electoral systems is emerging and should be used in designing new systems or reforming old ones. Regrettably, history demonstrates that elected officials have repeatedly manipulated systems for partisan advantage ... and have resisted the "intrusions" of scientific approaches to the design of electoral systems. Few voters realize the extent to which manipulation has profoundly effected electoral outcomes, sometimes transforming the votes of a minority into a majority in Parliament.

All too often the players of a nation's political game – its elected officials – are, at one and the same time, the referees of the game, and they change the rules to accommodate new situations. Imagine the public outcry were the game to be football! This is why independent commissions are needed, together with professionals trained in the emerging multi-disciplinary science of electoral systems, responsible for keeping abreast of all the new theoretical and technological developments in voting.

Erice, 23 September 2005

Editors' note: The present declaration, now known under the name of "The Erice Decalogue", has been unanimously signed by all the participants in the Erice Workshop. It has been written by Professor Michel Balinski with contributions by several invited speakers and the editorial assistance of Dr. Isabella Lari.

Power Indices Taking into Account Agents' Preferences

Fuad Aleskerov
State University 'Higher School of Economics'
and
Institute of Control Sciences, Russian Academy of Sciences

Abstract   A set of new power indices is introduced extending Banzhaf power index and taking into account agents' preferences to coalesce. An axiomatic characterization of intensity functions representing a desire of agents to coalesce is given. A set of axioms for new power indices is presented and discussed. An example of use of these indices for Russian parliament is given.

Keywords: Ordinal power index, Cardinal power index, intensity function, consistency of factions.

1.   Introduction

Power indices have become a very powerful instrument for study of electoral bodies and an institutional balance of power in these bodies (Brams, 1975; Felsenthal and Machover, 1998; Grofman and Scarrow, 1979; Herne and Nurmi, 1993; Laruelle and Valenciano, 2001; Leech, 2004). One of the main shortcomings mentioned almost in all publications on power indices is the fact that well-known indices do not take into account the preferences of agents (Felsenthal and Machover, 1998; Steunenberg et al., 1999). Indeed, in construction of those indices, e.g., Shapley-Shubik or Banzhaf power indices (Banzhaf, 1965; Shapley and Shubik, 1954), all agents are assumed to be able to coalesce. Moreover, none of those indices evaluates to which extent the agents are free in their wishes to create a coalition, how intensive are the connections inside one or another coalition[1].

Until recently the only index taking into account preferences of voters was that of Shapley – Owen (Shapley and Owen, 1989). However, the application

---

[1] First study on the coalition formation taking into account preferences of agents to coalesce was Dreze and Greenberg (1980). However, the problem of power distribution among agents in that study had not been considered.

of it to real data reveals some serious problems. They have been discussed in Barr and Passarelli (2004). That is why several attempts have been made to construct power indices do take into account preferences of voters to coalesce (Napel and Widgren, 2004, 2005).

In this article we try to construct another approach to define such indices. Consider an example. Let three parties A, B and C with 50, 49 and 1 sets, respectively, are presented in a parliament, and the voting rule is the simple majority one, i.e., 51 votes for. Then winning coalitions are $A + B$, $A + C$, $A + B + C$ and A is pivotal in all coalitions, B is pivotal in the first coalition and C is pivotal in the second one. (Normalized) Banzhaf power index[2] for these parties is equal to

$$\beta(A) = 3/5, \beta(B) = \beta(C) = 1/5.$$

Assume now that parties A and B never coalesce in pairwise coalition, i.e., coalition $A + B$ is impossible. Let us, however, assume that the coalition $A + B + C$ can be implemented, i.e., in the presence of 'moderator' C parties A and B can coalesce. Then the winning coalitions are $A + C$ and $A + B + C$, and A is pivotal in both coalitions while C is in one; B is pivotal in none of the winning coalitions. In this case $\beta(A) = 2/3$, $\beta(C) = 1/3$ and $\beta(B) = 0$, i.e., although B has almost half of the seats in the parliament, its power is equal to 0.

If A and B never coalesce even in the presence of a moderator C, then the only winning coalition is $A + C$, in which both parties are pivotal. Then, $\beta(A) = \beta(C) = 1/2$. Such situations are met in real political systems. For instance, Russian Communist Party in the second parliament (1997-2000) had had about 35% of seats, however, its power during that period was always almost equal to 0 (Aleskerov et al., 2003).

We introduce here two new types of indices based on the idea similar to Banzhaf power index, however, taking into account agents' preferences to coalesce. In the first type the information is used about agents' preferences over other agents. These preferences are assumed to be linear orders. Since these preferences may not be symmetric, the desire of agent 1 to coalesce with agent 2 can be different than the desire of agent 2 to coalesce with agent 1. These indices take into account in a different way such asymmetry of preferences. In the second type of power index the information about the intensity of prefer-

---

[2]Banzhaf power index is evaluated as

$$\beta(i) = \frac{b_i}{\sum_j b_j},$$

b_i is the number of winning coalitions in which agent i is pivotal, i.e., if agent i expels from the coalition it becomes a loosing one (Banzhaf, 1965). This form of Banzhaf index is called the normalized one.

ences is taken into account as well, i.e., we extend the former type of power index to cardinal information about agents' preferences.

The structure of the paper is as follows. Section 2 gives main notions. In Section 3 we define and discuss 'ordinal' power indices. In Section 4 cardinal indices are introduced. In Section 5 we evaluate power distribution of groups and factions in the Russian Parlament in 2000-2003 using some of new indices. Section 6 and 7 provides some axioms for the indices introduced. Section 8 cocludes.

2.   Main Notions

The set of agents is denoted as N, $N = \{1, ..., n\}$, $n > 1$. A coalition ω is a subset of N, $\omega \subseteq N$. We consider the situation when the decision of a body is made by voting procedure; agents who do not vote 'yes' vote against it, i.e., the abstention is not allowed.

Each agent $i \in N$ has a predefined number of votes, $v_i > 0, i = 1, \ldots, n$. It is assumed that a quota q is predetermined and as a decision making rule the voting with quota is used, i.e., the decision is made if the number of votes for it is not less than q,

$$\sum_i v_i \geq q.$$

The model describes a voting by simple and qualified majority, voting with veto (as in the Security Council of UN), etc.

A coalition ω is called winning if the sum of votes in the coalition is no less than q. An agent i is called pivotal in a coalition ω if the coalition $\omega \backslash \{i\}$ is a loosing one.

For such voting rule the set of all winning coalitions Ω possesses the following properties:

$$\begin{aligned}
\emptyset &\notin \Omega, \\
N &\in \Omega, \\
\omega &\in \Omega, \omega' \supseteq \omega \Longrightarrow \omega' \in \Omega.
\end{aligned}$$

Sometimes, one additional condition is applied as well

$$\omega \in \Omega \Longrightarrow N \backslash \omega \notin \Omega,$$

implying $q \geq \lceil \frac{n}{2} \rceil$, where $\lceil x \rceil$ is the smallest integer greater than or equal to x.

Next we introduce two types of indices, ordinal and cardinal. Both types are constructed on the following basis: the intensity of connection $f(i, \omega)$ of the agent with other members of ω is defined. Then for such agent i the value χ_i is evaluated as

$$\chi_i = \sum_\omega f(i,\omega),$$

i.e., the sum of intensities of connections of i over those coalitions in which i is pivotal.

Naturally, other functions instead of summation can be considered.

Then the power indices are constructed as

$$\alpha(i) = \frac{\chi_i}{\sum_j \chi_j}.$$

The very idea of the index α is the same as for Banzhaf index, with the difference that in Banzhaf index we evaluate the number of coalitions in which i is pivotal, i.e., in the definition of Banzhaf index χ_i is equal to 1, on the contrary, in our case χ_i is defined by the value of intensity function.

The main question is how to construct the intensity functions $f(i,\omega)$. Below we give two ways how to construct those functions.

Each agent i is assumed to have a linear order[3] P_i revealing her preferences over other agents in the sense that i prefers to coalesce with agent j rather than with agent k if P_i contains the pair (j,k). Obviously, P_i is defined on the Cartesian product $(N\backslash\{i\}) \times (N\backslash\{i\})$.

Since P_i is a linear order, the rank p_{ij} of the agent j in P_i can be defined. We assume that $p_{ij} = |N| - 1$ for the most preferable agent j in P_i.

The value p_{ij} shows how many agents less preferable than j are in P_i. For instance, if $N = \{A,B,C,D\}$ and $P_A : B \succ C \succ D$, then $p_{AB} = 3, p_{AC} = 2$ and $p_{AD} = 1$.

Using these ranks, one can construct different intensity functions.

A second way of construction of $f(i,\omega)$ is based on the idea that the values p_{ij} of connection of i with j are predetermined somehow. In general, it is not assumed $p_{ij} = p_{ji}$. Then the intensity function can be constructed as above.

Below we give six different ways how to construct $f(i,\omega)$ in ordinal case and sixteen ways of construction of cardinal function $f(i,\omega)$.

3.   Ordinal Indices

For each coalition ω and each agent i construct now an intensity $f(i,\omega)$ of connections in this coalition. In other words, f is a function which maps $N \times \Omega$ $(= (2^N\backslash\{\emptyset\})$ into $\mathcal{R}^1$, $f : N \times \Omega \to \mathcal{R}^1$. This very value is evaluated using the ranks of members of coalition. Several different ways to evaluate f using different information about agents' preferences are provided:

a) *Intensity of i's preferences.*

---

[3]i.e. irreflexive, transitive and connected binary relation. We often denote it as $\succ$.

In this form only preferences of i's agent over other agents are evaluated, i.e.,

$$f^+(i,\omega) = \sum_{j\in\omega} \frac{p_{ij}}{|\omega|};$$

b) *Intensity of preferences for i.* In this case we consider the sum of ranks of i given by other members of coalition ω.

$$f^-(i,\omega) = \sum_{j\in\omega} \frac{p_{ji}}{|\omega|};$$

c) *Average intensity with respect to i's agent*

$$f(i,\omega) = \frac{f^+(i,\omega) + f^-(i,\omega)}{2};$$

d) *Total positive average intensity.*

Consider any coalition ω of size $k \leq n$. Without loss of generality one can put $\omega = \{1, \ldots, k\}$. Then consider $f^+(i,\omega)$ for each i and construct

$$f^+(\omega) = \frac{\displaystyle\sum_{i\in\omega} f^+(i,\omega)}{|\omega|};$$

e) *Total negative average intensity* is defined similarly by the formula

$$f^-(\omega) = \frac{\displaystyle\sum_{i\in\omega} f^-(i,\omega)}{|\omega|};$$

f) *Total average intensity* is defined as

$$f(\omega) = \frac{\displaystyle\sum_{i\in\omega} f(i,\omega)}{|\omega|}.$$

It is worth emphasizing here that the intensities d) – f) do not depend on agent i, i.e., for any agent i in the following calculation of power indices we assume that for any i in the coalition ω the corresponding intensity is the same.

Consider now several examples.

Example 1. Let $n = 3, N = \{A, B, C\}, v(A) = v(B) = v(C) = 33, q = 50$. Consider two preference profiles given in Tables 1 and 2.

For both preference profiles there are three winning coalitions in which agents are pivotal. These coalitions are $A + B, A + C$ and $B + C$.

Table 1.   First preference profile

P_A	P_B	P_C
C	C	A
B	A	B

Table 2.   Second preference profile

P_A	P_B	P_C
B	C	A
C	A	B

Let us calculate the functions f as above for each agent in each winning coalition. The preferences from Tables 1 and 2 can be re-written in the matrix form as

$$\|p_{ij}\| = \begin{array}{c} \\ A \\ B \\ C \end{array} \begin{array}{ccc} A & B & C \\ \left(\begin{array}{ccc} 0 & 1 & 2 \\ 1 & 0 & 2 \\ 2 & 1 & 0 \end{array} \right) \end{array}$$

$$\|p_{ij}\| = \begin{array}{c} \\ A \\ B \\ C \end{array} \begin{array}{ccc} A & B & C \\ \left(\begin{array}{ccc} 0 & 2 & 1 \\ 1 & 0 & 2 \\ 2 & 1 & 0 \end{array} \right) \end{array}$$

Now, for the profile given in Table 1 one can calculate the values of intensities a)–f) obtained by each agent i in each winning coalition ω. These values for the first preference profile are given in Table 3 and for the second one – in Table 4.

Using these intensity functions one can define now the corresponding power indices $\alpha(i)$. Let i be a pivotal agent in a winning coalition ω. Denote by χ_i the number equal to the value of the intensity function for a given coalition ω and agent i. Then the power index is defined as follows

$$\alpha(i) = \frac{\displaystyle\sum_{\substack{\omega \\ i \text{ is pivotal in } \omega}} \chi_i}{\displaystyle\sum_{j \in N} \sum_{\substack{\omega \\ j \text{ is pivotal in } \omega}} \chi_j}$$

Table 3. Intensity values for the first preference profile

	$f^+(i,\omega)$			$f^-(i,\omega)$			$f(i,\omega)$		
	A	B	C	A	B	C	A	B	C
$A+B$	1/2	1/2	-	1/2	1/2	-	1/2	1/2	-
$A+C$	1	-	1	1	-	1	1	-	1
$B+C$	-	1	1/2	-	1/2	1	-	3/4	3/4
	$f^+(i,\omega)$			$f^-(i,\omega)$			$f(i,\omega)$		
	A	B	C	A	B	C	A	B	C
$A+B$	1/2	1/2	-	1/2	1/2	-	1/2	1/2	-
$A+C$	1	-	1	1	-	1	1	-	1
$B+C$	-	3/4	3/4	-	3/4	3/4	-	3/4	3/4

Table 4. Intensity values for the second preference profile

	$f^+(i,\omega)$			$f^-(i,\omega)$			$f(i,\omega)$		
	A	B	C	A	B	C	A	B	C
$A+B$	1	1/2	-	1/2	1	-	3/4	3/4	-
$A+C$	1/2	-	1	1	-	1/2	3/4	-	3/4
$B+C$	-	1	1/2	-	1/2	1	-	3/4	3/4
	$f^+(i,\omega)$			$f^-(i,\omega)$			$f(i,\omega)$		
	A	B	C	A	B	C	A	B	C
$A+B$	3/4	3/4	-	3/4	3/4	-	3/4	3/4	-
$A+C$	3/4	-	3/4	3/4	-	3/4	3/4	-	3/4
$B+C$	-	3/4	3/4	-	3/4	3/4	-	3/4	3/4

Table 5. Power indices values

	First profile (Table 1)			Second profile (Table 2)		
	A	B	C	A	B	C
α_1	1/3	1/3	1/3	1/3	1/3	1/3
α_2	1/3	2/9	4/9	1/3	1/3	1/3
α_3	1/3	5/18	7/18	1/3	1/3	1/3
α_4	1/3	5/18	7/18	1/3	1/3	1/3
α_5	1/3	5/18	7/18	1/3	1/3	1/3
β	1/3	5/18	7/18	1/3	1/3	1/3

As we already mentioned this index is similar to the Banzhaf index. The difference is that in the Banzhaf index χ_i is equal to 1, in the case under study χ_i represents some intensity value.

The indices $\alpha(i)$ will be denoted by $\alpha_1(i), \ldots, \alpha_6(i)$.

Let us evaluate now the values $\alpha_1(\cdot) - \alpha_6(i)$ for all agents for the preference profile from Table 1.

The agent A (as well as agents B and C) is pivotal in two coalitions; the sum of the values $f^+(i, \omega)$ for each i is equal to 3/2. Then

$$\alpha_1 = \frac{3/2}{3/2 + 3/2 + 3/2} = \frac{1}{3} = \alpha_1(B) = \alpha_1(C).$$

The value $\alpha_2(\cdot)$ is evaluated differently. The sum of values $f^-(i, \omega)$ from Table 3 for all i and ω is equal to 9/2. However, for A $\sum_\omega f(A, \omega) = 3/2$, $\sum_\omega f(B, \omega) = 1$ and $\sum_\omega f(C, \omega) = 2$. Then $\alpha_2(A) = \frac{3}{9} = \frac{1}{3}$; $\alpha_2(B) = \frac{2}{9}$ and $\alpha_2(C) = \frac{4}{9}$.

The values of the indices $\alpha_1(\cdot) - \alpha_6(i)$ for both preference profiles are given in Table 5 as well as the values of Banzhaf index β.

Consider now another example.

Example 2. Let $N = \{A, B, C, D, E\}$, each agent has one vote, $q = 3$ and the preferences of agents are given in Table 6. The values of indices $\alpha_2(\cdot) - \alpha_4(i)$ are given in Table 7.

Note that α_1 is equal to the Banzhaf index, which for this case gives $\forall i \in N$ $\beta(i) = 1/5$.

Example 3. Consider the case when 3 parties A, B and C have 50, 49 and 1 seats, respectively. Assume that decision making rule is simple majority, i.e. 51 votes. Then the winning coalitions are A+B, A+C and A+B+C. Note that

Table 6. Preferences of agents for $N = \{A, B, C, D, E\}$

P_A	P_B	P_C	P_D	P_E	rank
B	A	D	A	B	4
C	C	A	B	A	3
D	D	B	C	D	2
E	E	E	E	C	1

Table 7. The values of the indices $\alpha_2 - \alpha_4$ for Example 2.

	A	B	C	D	E
α_2	0.28	0.26	0.18	0.2	0.008
α_3	0.24	0.23	0.19	0.2	0.14
α_4	0.22	0.21	0.2	0.2	0.17

A is pivotal in all three coalitions, B and C are pivotal in one coalition each. Then $\beta(A) = 3/5$, $\beta(B) = \beta(C) = 1/5$.

Consider now the case with the preferences of agents given below: $P_A; C \succ B; P_B : C \succ A$ and $P_C : A \succ B$.

Then the values of α_1 and α_2 (constructed by $f^+(i, \omega)$ and $f^-(i, \omega)$) are as follows

$$\alpha_1(A) = 5/12, \quad \alpha_1(B) = 1/4, \quad \alpha_1(C) = 1/3,$$
$$\alpha_2(A) = 5/12 \quad \alpha_2(B) = 7/36 \quad \alpha_2(C) = 7/18.$$

Consider another preference profile: $P'_A : C \succ B$, $P'_B : C \succ A$ and $P'_C : B \succ A$, i.e., the only agent C changes her preferences. Then one can easily evaluate $\alpha'_1(A) = 5/11$, $\alpha'_1(B) = 3/11$, $\alpha'_1(C) = 3/11$, $\alpha'_2(A) = 10/33$, $\alpha'_2(B) = 3/11$, $\alpha'_2(C) = 14/33$.

4. Cardinal Indices

Assume now that the desire of party i to coalesce with party j is given as real number p_{ij}, $\sum_j p_{ij} = 1$, $i, j = 1, \ldots, n$. In general, it is not assumed that $p_{ij} = p_{ji}$.

One can call the value p_{ij} as an intensity of connection of i with j. It may be interpreted as, for instance, a probability for i to form a coalition with j.

We define now several intensity functions

a) average intensity of i's connection with other members of coalition ω

$$f^+(i,\omega) = \frac{\sum\limits_{j \in \omega} p_{ij}}{|\omega|};$$

b) average intensity of connection of other members of coalition with i

$$f^-(i,\omega) = \frac{\sum\limits_{j \in \omega} p_{ji}}{|\omega|};$$

c) average intensity for i

$$f(i,\omega) = \frac{1}{2}\left(f^+(i,\omega) + f^-(i,\omega)\right);$$

d) average positive intensity in ω

$$f^+(i,\omega) = \frac{\sum\limits_{i \in \omega} f^+(i,\omega)}{|\omega|},$$

e) average negative intensity in ω

$$f^-(i,\omega) = \frac{\sum\limits_{i \in \omega} f^-(i,\omega)}{|\omega|},$$

f) average intensity in ω

$$f(\omega) = \frac{\sum\limits_{i \in \omega} f(i,\omega)}{|\omega|},$$

In contrast to ordinal case now we can introduce several new intensity functions:

g) minimal intensity of i's connections

$$f^+_{\min}(i,\omega) = \min_j p_{ij};$$

h) maximal intensity of i's connections

$$f^+_{\max}(i,\omega) = \max_j p_{ij};$$

i) maximal fluctuation of i's connections

$$f_{mf}(i,\omega) = \frac{1}{2}\left(\min_j p_{ij} + \max_j p_{ij}\right);$$

j) minimal intensity of connections of other agents in ω with i

$$f_{\min}^-(i,\omega) = \min_j p_{ji}$$

k) maximal intensity of connections of other agents ω in with i

$$f_{\max}^-(i,\omega) = \max_j p_{ji}$$

l) s-mean intensity of i's connections with other agents in ω

$$f_{sm}^+(i,\omega) = \frac{1}{|\omega|}\sqrt[s]{\sum_j p_{ij}^s};$$

m) s-mean intensity of connections of other agents ω in with i

$$f_{sm}^+(i,\omega) = \frac{1}{|\omega|}\sqrt[s]{\sum_j p_{ji}^s};$$

n) max min intensity

$$f_{\max\min}(\omega) = \max_i \min_j p_{ij};$$

o) min max intensity

$$f_{\min\max}(\omega) = \min_i \max_j p_{ji};$$

p) maximal fluctuation

$$f_{mf}(\omega) = \frac{1}{2}\left(f_{\max\min}(\omega) + f_{\min\max}(\omega)\right).$$

Note that the intensity functions in the cases d)–f), n)–p) do not depend on agent herself but only on coalition ω.

Now the corresponding power indices can be defined as above, i.e.,

$$\alpha^{\text{card}}(i) = \frac{\displaystyle\sum_{\substack{\omega \text{ is winning} \\ i \text{ is pivotal in }\omega}} \chi_i}{\displaystyle\sum_{j\in N}\sum_{\substack{\omega \text{ is winning} \\ j \text{ is pivotal in }\omega}} \chi_i(\omega)},$$

where χ_i is one of the above intensity functions.

Example 4. Let $N = \{A, B, C, D\}$, each voter has only one vote, the quota is equal to $q = 3$, and the matrix $\|p_{ij}\|$ is given in Table 8. In Table 9 the power indices are given for the cases a), b), e), h).

Table 8. Matrix $\|p_{ij}\|$ for Example 3

	A	B	C	D
A		0.7	0.2	0.1
B	0.3		0.5	0.2
C	0.1	0.7		0.2
D	0.7	0.2	0.1	

Table 9. Some cardinal indices for Example 3

	A	B	C	D
$\alpha_{a)}$	0.25	0.25	0.25	0.25
$\alpha_{b)}$	0.27	0.40	0.20	0.13
$\alpha_{c)}$	0.25	0.27	0.24	0.23
$\alpha_{h)}$	0.25	0.25	0.25	0.25

5. Evaluation for Russian Parliament

We will study now a distribution of power among factions in the third Russian Parliament (1999-2003) using these new indices. The matrix $\|p_{ij}\|$ is constructed using the consistency index; the latter (the index of consistency of positions of two groups) is constructed as

$$c(q_1, q_2) = 1 - \frac{|q_1 - q_2|}{\max(q_1, 1 - q_1, q_2, 1 - q_2)},$$

where q_1 and q_2 be the share of "ay" votes in two groups of MPs (Aleserkov et al., 2003).

We consider the value of consistency index as the value of intensity of connections between agents i and j. Then we are in cardinal framework, and one can use one of the indices introduced in the previous section.

On Fig. 1 the values of $\alpha_{a)}$ index are given for the Russian Parliament from 2000 to 2003 on the monthly basis. It can be readily seen that index α gives lower values for Communist Party (sometimes up to 3%) and higher values for Edinstvo (up to 1%). It is interesting to note that Liberal-Democrats (Jirinovski's Party) had had almost equal values by both indices, which corresponds to the well-known flexibility of that party position.

Let us note that different ways to use the index α are possible. For instance, following the approach from Aleskerov et al. (2003), we may assume that if the consistency value for two factions is less than some threshold value δ, then parties do not coalesce, i.e., $p_{ij} = 0$.

Figs. 2 and 3 give power distribution for the factions in the Russian Parliament for the same period calculated on the basis of factions coordinates on a political map. On that map each faction at each period is characterized by two coordinates – the level to which extent it is liberal or state oriented and the level of support of the president (pro-reforms or anti-reforms) (Aleskerov et al., 2005).

Having these two coordinates, we calculate the distance on the map between the positions of two factions. Then it is possible to construct a measure τ_{ij} – intensity of connections among factions i and j – as

$$\tau_{ij} = \frac{1}{\sqrt{2}} \left(\frac{1 + \sqrt{2}}{1 + d_{ij}} - 1 \right),$$

where d_{ij} is the Euclidean distance between positions of factions i and j on political map.

It can be easily seen that $\tau_{ij} = 0$ if $d_{ij} = \sqrt{2}$ (the maximal distance on the map), and $\tau_{ij} = 1$ if $d_{ij} = 0$ (i.e., when positions of two factions coincide). Using the values τ_{ij} for two factions and consider them as a measure of preference to coalesce, one can calculate the cardinal indices introduced above, in particular, the index for the case a). These very evaluations are given on Figs. 2 and 3 for five main parties in Russian parliament during the period 2000-2003.

6.   Axiomatic Construction of a Cardinal Intensity Function

Now we will try to axiomatize a construction of cardinal intensity function.

First, we define an intensity function depending on intensities p_{ij} of connections of i with other members of coalition ω, i.e., if $\omega = \{1, \ldots, m\}$, $m \leq n$,

$$f(i, \omega) = f_i (p_{11}, \ldots, p_{1m}, p_{21}, \ldots, p_{2m}, \ldots, p_{i1}, \ldots, p_{im}, \ldots, p_{mm}).$$

As it is seen, the intensity function for i depends not only of i's connections with other members of coalition, but depends also of connections of other members among themselves. We can consider, for instance, the case when the intensity of agent i to join a coalition ω depends on the average intensity of connections between members of ω, say, the intensity can be low if that average intensity is below some threshold.

However, we will restrict this function in a way which is similar to independence of irrelevant alternatives (Arrow, 1963): $f(i, \omega)$ will depend on connections of agent i with other members of coalition ω only, i.e.,

$$f(i, \omega) = f_i (p_{i1}, \ldots, p_{im}).$$

For the sake of simplicity we put $p_{ij}^\omega \geq 0$ for all i, j and $\forall i \sum_{j \in \omega} p_{ij}^\omega = 1$.

I would like to emphasize that in this formulation the sum of p_{ij}^ω is equal to 1 in each ω, i.e., now connections are defined by $2^N - 1$ matrices $\|p_{ij}^\omega\|$ for each coalition ω.

Consider several axioms which reasonable function $f(i, \omega)$ should satisfy to.

Axiom 1. For any $m -$ tuple of values $(p_{i1}, \ldots, p_{im})$ there exist a function $f(i, \omega)$ such that $0 \leq f(i, \omega) \leq 1$, f is continuous differentiable function of each of its arguments.

Axiom 2. If $p_{ij} = 0$ for any j, then $f(i, \omega) = 0$.

Axiom 3. (Monotonicity). A value of $f(i, \omega)$ increases if any value p_{ij} increases, and a value of $f(i, \omega)$ decreases if p_{ij} decreases. Moreover, equal changes in intensities p_{ij} lead to equal changes of $f(i, \omega)$. This means that

$$\frac{\partial f_i}{\partial p_{ij}} = \mu_i \quad \text{for any } j,$$

and

$$\frac{\partial f_i}{\partial p_{lj}} = 0 \quad \text{for any } l \neq i.$$

Then the following theorem holds

Theorem. An intensity function $f(i, \omega)$ satisfies Axioms 1–3 iff it is represented in the form

$$f(i, \omega) = \frac{\sum_j p_{ij}}{|\omega|}.$$

Proof is a re-formulation of the proof of the theorem from Intriligator (1973) given in the framework of probabilistic social choice and hence is omitted.

An axiomatic characterization of other types of intensity functions is still an open problem.

7.   Axioms for Power Indices

We introduce several axioms, which any reasonable power index should satisfy to.

First, we call a voting situation a four-tuple $[N, q, v, \vec{P}]$, where N is a set of agents, $|N| = n$, $n > 1$, q is a quota, $v = (v_1, \ldots, v_n)$ is a set of votes which agents possess, $\vec{P}$ is a preference profile, where each agent $i \in N$ has a preference (linear order) P_i over $N \backslash \{i\}$ or preference matrix $\|p_{ij}\|$.

Axiom 1. Under a given quota rule for any agent $i \in N$ there exists a preference profile $\vec{P}$ such that $\alpha(i) > 0$.

In words, for no agent it is known in advance, independently of agents' preferences, that her power is equal to 0.

Axiom 2. Consider two voting situations $[N, q, v, \vec{P}]$ and $[N, q, v', \vec{P}]$. Let $\exists A \in N$ such that $v'(A) \geq v(A)$, and $\forall B \in N$, $v'(B) = v(B)$. Then, $\alpha'(A) \geq \alpha(A)$.

Assume that for a given distribution of votes and a given preference profile we evaluate power distribution among agents. Then we increase the number of votes for a given agent A, keeping the votes of other agents unchanged. Then Axiom 2 states that voting power of A in new situation should not be less than before.

Axiom 3. (Symmetry) Let η be a one-to-one correspondence of N to N. Then

$$\eta(\alpha_1, \ldots, \alpha_n) = (\alpha_{\eta(1)}, \ldots, \alpha_{\eta(n)}).$$

Axiom 3 states that power of agents does not depend of their names, i.e., the procedure of evaluation of power distribution must treat agents in a similar way.

Axiom 4. Let $i \in N$ be pivotal in no winning coalition ω. Then, $\alpha(i) = 0$.

It is usual axiom in voting power models (in fact, in game – theoretic models, see Shapley and Shubik (1954)): a dummy player has power equal to 0.

Axiom 5'. First Monotonicity Axiom (FMA). Consider two voting situations $[N, q, v, \vec{P}]$ and $[N, q, v, \vec{P'}]$. Let for some i and any $k \neq i$ $P_k = P'_k$ holds. Let additionally for some $p'_{ij} > p_{ij}$ holds. Then, $\alpha'(j) \geq \alpha(j)$.

This axiom can be explained in a simple way: all preferences except i's are the same in two profiles; in i'th preference the evaluation of j is higher in new profile than in the old one. Then in new voting situation (with $\vec{P'}$) the power of j should not be less than before.

Axiom 5''. Second Monotonicity Axiom (SMA). Consider two voting situations $[N, q, v, \vec{P}]$ and $[N, q, v, \vec{P'}]$. Let for two agents i and j $\alpha(i) \geq \alpha(j)$ holds, where $\alpha(i)$ is the voting power of i in the first voting situation. Let $\vec{P'}$ is such that for any $k \neq l$ $P_k = P'_k$ holds, and in the preferences of l's agent

$$p'_{li} - p'_{lj} > p_{li} - p_{lj}$$

holds.

Then $\alpha'(i) \geq \alpha(j)$ (weak version of SMA) or $\alpha'(i) > \alpha(j)$ (strong version of SMA), where $\alpha'(i)$ is the voting power of i with respect to second voting situation.

In words, assume that the power of i is not less than the power of j with respect to first voting situation. Let $\vec{P'}$ is such that for any agent but ℓ her new preferences coincide with old ones, and in l's preference the relative position of i with respect to j is higher in P'_l than in P_l. Then the voting power of i

should be not less than that of j in new voting situation (weak version) or even must be greater than that of j (strong version).

Axiom 6. Let $\vec{P}'$ be an intensity matrix such that $p'_{ij} = kp_{ij}$ for every $i, j = 1, \ldots, n$. Then $\alpha'(i) = \alpha(i)$ where α' is the power vector obtained from $\vec{P}'$.

Axiom 6 deals with cardinal power indices. It says that voting power of agents does not change under the transformation of scale of intensities in the form

$$p'_{ij} = kp_{ij},$$

i.e., when intensities multiply to the same constant k.

It is possible to formulate axioms similar to those from Section 5 and prove a theorem similar to the given above but for α–indices. However, it will be interesting to analyze how the axioms from this Section provide an axiomatic characterization of α indices.

8.   Concluding Remarks

We have considered three ways to construct power indices taking into account voters' preferences to coalesce. The first one is based on the consistency index showing to which extent two groups of voters (party factions) vote in a similar way. The values of consistency index define the possibility of these groups to coalesce. Then the Banzhaf index is defined on the set of admissible coalitions only.

The second way uses the functions defining the intensity of factions to coalesce being based on the intensity to coalesce of individual faction. We have defined six ordinal intensity indices and sixteen cardinal ones. For a simplest cardinal intensity index the corresponding axioms are introduced and the characterization theorem is proved.

Then the power index is defined in a way similar to Banzhaf index – instead of calculating number of coalitions in which faction is pivotal we calculate an intensity of faction to coalesce in the coalitions in which it is pivotal.

Finally, we define an intensity function as a function of distance using the coordinates of factions on the political map. The latter is constructed using data of real voting in a parliament (see, for instance, Aleskerov et al. (2005)).

Then using this intensity of faction one can calculate one of the power indices defined above for a cardinal case.

Acknowledgments

I am very thankful to the participants of the Workshop on Constitutional and Scientific Quandaries (International Center on Economic Research, Turin, Italy, June 20-23, 2005), Workshop on Voting Power and Procedures (Warwick

University, UK, July 20-23, 2005), Workshop on Mathematics and Democracy (Erice, Sicily, Italy, September 19-23, 2005) and specially to Professors Norman Schofield, Kenneth Shepsle, Bruno Simeone, John Ferejohn, Moshe Machover, Friedrich Pukelsheim, Federico Valenciano and anonymous referee for helpful comments.

Mrs. Natalya Andryshina and Ms. Anna Sokolova helped me a lot to prepare the text. Mr. Vyacheslav Yakuba prepared software to evaluate the indices.

This work was also supported by the grant of the State University 'Higher School of Economics' (grant no. 285/10-04). Professors Vladimir Avtonomov, Yaroslav Kuz'minov, Andrey Yakovlev and Evgeniy Yasin had supported the project from its very beginning. Finally, this work is partially supported by the grant no. 05-01-00188 of Russian Foundation of Basic Research.

All these help and support are gratefully acknowledged.

References

Aleskerov F., Blagoveschenskiy N., Satarov G., Sokolova A., Yakuba V. "Evaluation of Power of Groups and Factions in the Russian Parliament (1994-2003)", WP7/2003/01, Moscow: State University "High School of Economics", 2003 (in Russian).

Aleskerov F., Blagoveschensky N., Konstantinov M., Satarov G., Yakuba V. "A Balancedness of the 3d State Duma of Russian Federation Evaluated by the Use of Cluster-analysis", WP7/2005/04, Moscow: State University "High School of Economics", 2005 (in Russian)

Arrow K.J. Social Choice and Individual Values, – 2nd ed., New Haven: Yale University Press, 1963.

Banzhaf J. F., "Weighted Voting Doesn't Work: A Mathematical Analysis", Rutgers Law Review, 1965, v.19, 317-343.

Barr J. and F. Passarelli, "Who has the power in the EU?", Working Papers Rutgers University, Newark, no. 2004-005, Department of Economics, Rutgers University, Newark, 2004.

http://www.rutgers-newark.rutgers.edu/econnwk/workingpapers/2004-005.pdf

Brams S. "Game Theory and Politics", The Free Press, New York, 1975.

Dreze J.H., Greenberg J. "Hedonic Coalitions: Optimality and Stability", Econometrica, v.48, no.4, 1980, 987-1003.

Felsenthal D., Machover M. "The Measurment of Voting Power: Theory and Practices, Problems and Paradoxes", Edgar Elgar Publishing House, 1998.

Grofman B. and H. Scarrow "Ianucci and Its Aftermath: The Application of Banzhaf index to Weighted Voting in the State of New York", Brams, S., Schotter A. and G. Schwodiatuer (eds.) Applied Game Theory, 1979.

Herne K. and H. Nurmi "A Priori Distribution of Power in the EU Council of Ministers and the European Parliament", Scandinavian Journal of Poilitical Studies, 1993, v.16, 269-284.

Intriligator M.D. "A Probabilistic Model of Social Choice", The Review of Economic Studies, v.XL(4), 1973, 553-560.

Laruelle A., Valenciano F. "Shapley-Shubik and Banzhaf Indices Revisited", Mathematics of Operations Research, v.26, 2001, 89-104.

Leech, D. "Voting Power in the Governance of the International Monetary Fund", Annals of Operations Research, 2002, v.109, 375-397.

Napel S. and M. Widgren "Power Measurement as Sensitivity Analysis – a Unified Approach", Journal of Theoretical Politics 16, 2004, 517-538.

Napel S. and M. Widgren "The Possibility of Preference Based Power Index", Journal of Theoretical Politics 17, 2005, 377-387.

Shapley, L.S., and G. Owen "Optimal location of candidates in ideological space", International Journal of Game Theory. 1989, no.1, 125-142.

Shapley, L.S., and M. Shubik "A Method for Evaluting the Distribution of Power in a Committee System", American Political Science Review, 1954, v. 48, 787-792.

B. Steunenberg, D. Schmidtchen, Chr. Coboldt, "Strategic Power in the European Union", Journal of Theoretical Politics 11, 1999, 339-366.

The Sunfish Against the Octopus:
Opposing Compactness to Gerrymandering

Nicola Apollonio[1], Ronald I. Becker[2], Isabella Lari[1], Federica Ricca[3], Bruno Simeone[1]

[1]Dip. Statistica, Probabilità e Statistiche Applicate, Università di Roma "La Sapienza"

[2]Dep. of Mathematics and Applied Mathematics, University of Cape Town

[3]Dip. Sistemi e Istituzioni per l'Economia, Università de L'Aquila

Abstract   Gerrymandering - the artful and partisan manipulation of electoral districts - is a well known pathology of electoral systems, especially majoritarian ones. In this paper, we try to give theoretical and experimental answers to the following questions: 1) How much biased can the assignment of seats be under the effect of gerrymandering? 2) How effective is compactness as a remedy against gerrymandering? Accordingly, the paper is divided into two parts. In the first one, a highly stylized combinatorial model of gerrymandering is studied; in the second one, a more realistic multiobjective graph-partitioning model is adopted and local search techniques are exploited in order to find satisfactory district designs. In a nutshell, our results for the theoretical model mean that gerrymandering is as bad as one can think of and that compactness is as good as one can think of. These conclusions are confirmed to a large extent by the experimental results obtained with the latter model on some medium-large real-life test problems.

Keywords: Gerrymandering, partition, graph coloring.

1.   Introduction

Gerrymandering - the partisan manipulation of electoral district boundaries - has plagued modern democracies since their early times. Far from being defeated, it keeps displaying its perverse effects even at present (Balinski, 2004). It was only with the rise of the electronic computer that researchers started thinking about neutral and rational procedures for political districting. Its nature as a multicriteria decision problem was soon recognized. Suppose that the territory is subdivided into elementary administrative units (counties, townships, wards,..). The most commonly adopted districting criteria are the following: integrity (no unit may be split between two or more districts); contiguity (the units within the same district should be geographically contiguous);

population equality (the district populations should be equal or nearly equal, especially in majoritarian systems); compactness (each district should be compact, that is, "closely and neatly packed together" (Oxford Dictionary)); conformity to administrative boundaries (the electoral district boundaries should not cross other administrative boundaries, such as those of regions, provinces, local or minority communities). Among these criteria, compactness stands as a powerful weapon against gerrymandering, since it bans indented or elongated districts: a sunfish-shaped district is deemed to be compact, while an octopus-shaped or an eel-shaped one is not.

The present paper deals with the following two basic problems:

1) How bad can gerrymandering be?

2) How effective is compactness in preventing gerrymandering?

We shall give both theoretical and experimental answers to these two problems. Accordingly, our paper is divided into two parts. In the first one, an idealized combinatorial model is investigated; in the second part, a more realistic and flexible multicriteria graph-theoretic model is adopted, and computational results are presented for some medium to large real-life test problems.

2. A Combinatorial Gerrymandering Model

As a motivation for the present section we mention a striking artificial example of gerrymandering given by Dixon and Plischke (1950). Suppose that only two parties P and C compete under a first-past-the-post system and that, as in Figure 1, the territory is divided into elementary units having the same population with an homogeneous electoral behavior, that is, the whole population of an elementary unit votes for the same party. If the district map of Figure 1 (a) is adopted, party C wins in 8 districts out of 9; however, if the alternative district map of Figure 1(b) is adopted, party C wins only in 2 districts out of 9, so the outcome is drastically reversed.

A careful look at Figure 1 gives us a clue about an effective strategy for maximizing the number of districts won by either party: the districts should be designed so that every win should be close and every loss should be sweeping.

In this section we shall consider an idealized graph-theoretic formulation that captures the essence of the artificial example by Dixon and Plischke. Given a territory composed by territorial units, define the following integers:

- n is the number of territorial units;

- p is the number of districts;

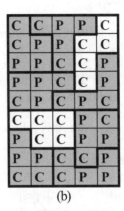

(a)   (b)

Fig. 1.   Example by Dixon and Plischke: (a) Party P wins 1 seat and party C wins 8; (b) Party P wins 7 seat and party C wins 2.

- s is the common district size (number of territorial units in each district).

Clearly, the three parameters n, p, s must satisfy the relation $n = ps$.

We model the territory as an undirected graph $G = (V, E)$ with $|V| = n$, where the vertices represent territorial units and the edges represent adjacency between territorial units.

A *connected partition* of G is a partition of its set of vertices V such that each component induces a connected subgraph of G.

A *district design* is a connected partition of the graph into p components or *districts* of the same size. Notice that this definition takes into account the criteria of integrity, contiguity and population equality. If at least one such partition exists, the graph is said to be p-equipartitionable. Checking such property is not easy: in fact, Frieze and Dyer (1985) proved its NP-completeness even for bipartite graphs. We assume that G is p-equipartitionable.

A *vote outcome* is a bicoloring of the vertices that assigns to each vertex either the color blue or the color red: this means that all voters in the corresponding unit vote for the same party, *blue* or *red*, respectively. A vote outcome is *balanced* if the number of blue vertices is equal to the number of red ones.

A balanced vote outcome corresponds to a situation in which the electoral population is perfectly split between two parties.

From now on, except for the last section, we shall consider only balanced vote outcomes. We shall also make the following assumptions on the integers n, s, and p:

- n is even: this is a necessary condition for the existence of balanced vote outcomes;

- s is odd and greater or equal to 3: this assumption forbids trivial cases and ties between the two parties;

- p is even: this follows from the relation $n = ps$.

If in a district D the number of blue vertices is greater than the number of red ones, we will say that D is a *blue district*. In a similar way we can define a *red district*. We will denote by Π the set of all district designs and by Ω the set of all possible balanced vote outcomes.

We define an *electoral competition* to be a pair (ω, π) such that $\omega \in \Omega$ and $\pi \in \Pi$. The functions $b(\omega, \pi)$ and $r(\omega, \pi)$, compute the number of blue and red districts, respectively, resulting from the electoral competition (ω, π). Let

$$B(G) = \max_{\omega \in \Omega, \pi \in \Pi} b(\omega, \pi)$$

be the maximum number of blue districts for all the electoral competitions $(\omega, \pi) \in \Omega \times \Pi$. In a similar way we can define $R(G)$ with respect to $r(\omega, \pi)$.

PROPERTY 1 *Since, for any bicoloring, it is possible to switch the colors of the vertices so that the red vertices become the blue vertices and viceversa, any property related to the blue party that does not explicitly depend on any given bicoloring must hold for the red party also. In particular we have that $B(G) = R(G)$.*

By this property we can define the function

$$W(G) = B(G) = R(G).$$

Moreover the results that we will provide for the blue party hold also for the red one.

Given an electoral competition $(\omega, \pi) \in \Omega \times \Pi$, for any district k, $k = 1, ..., p$, let

- b_k = number of blue vertices in district k,

- r_k = number of red vertices in district k.

PROPERTY 2 *Given a p-equipartitionable graph G, for any $(\omega, \pi) \in \Omega \times \Pi$ the following inequality holds:*

$$b(\omega, \pi) \le \lfloor n/(s+1) \rfloor.$$

Proof. Given an electoral competition $(\omega, \pi) \in \Omega \times \Pi$, for each district k let b_k and r_k be defined as above. Since ω is balanced, we may assume:

$$\sum_{k=1,\dots,p} (b_k - r_k) = 0.$$

Hence:

$$0 = \sum_{k=1,\dots,p} (b_k - r_k) = \sum_{k:b_k > r_k} (b_k - r_k) + \sum_{k:b_k < r_k} (b_k - r_k)$$

$$\ge b(\omega, \pi) - s(p - b(\omega, \pi)) = (s+1)b(\omega, \pi) - sp$$

Since $n = ps$ and $b(\omega, \pi)$ is a natural number we obtain:

$$b(\omega, \pi) \le \lfloor n/(s+1) \rfloor.$$

∎

COROLLARY 1 *If G is p-equipartitionable, then $W(G) = \lfloor n/(s+1) \rfloor$.*

Proof. Let $\pi \in \Pi$ be any district design. It is possible to color the vertices of the graph G in such a way that $\lfloor n/(s+1) \rfloor$ districts have at least $(s+1)/2$ blue vertices. In fact, in any balanced vote outcome, the number of blue vertices is $n/2$ and:

$$\frac{s+1}{2} \left\lfloor \frac{n}{s+1} \right\rfloor \le \frac{n}{2}.$$

Since a district with $(s+1)/2$ blue vertices is blue, we obtain a vote outcome with at least $\lfloor n/(s+1) \rfloor$ blue districts. But, by Proposition 2, this is an upper bound for the number of blue districts, hence $W(G) = \lfloor n/(s+1) \rfloor$. ∎

COROLLARY 2 *If G is p-equipartitionable, and $p = q(s+1) + r$ with $1 \le r \le s+1$ then $W(G) = qs + r - 1$[1].*

Proof. From Corollary 1 we have:

$$W(G) = \left\lfloor \frac{n}{s+1} \right\rfloor = qs + \left\lfloor \frac{rs}{s+1} \right\rfloor.$$

---

[1] Notice that q and r might not coincide with the quotient and the remainder, respectively, of the division of p by $s+1$.

Since $r \leq s + 1$,

$$\left\lfloor \frac{rs}{s+1} \right\rfloor = \left\lfloor r - \frac{r}{s+1} \right\rfloor = r - 1,$$

hence

$$W(G) = qs + r - 1.$$

∎

Given a bicoloring $\omega \in \Omega$ and a partition $\pi \in \Pi$, we say that a district is (*blue*) *edgy* if it contains $(s+1)/2$ blue vertices and $(s-1)/2$ red vertices, while we will say that a district is (*blue*) *sweeping* if all its vertices are blue. Moreover we say that a district design π is (*blue*) *extremal* if the number of blue districts $b(\omega, \pi)$ is equal to its upper bound $\lfloor n/(s+1) \rfloor$. Similar concepts can be introduced for the red party.

REMARK 3 *If* $p \leq s+1$, *each blue extremal partition has* $p-1$ *blue districts and one red district.*

We are especially interested in the following optimization problem:

$$GAP(G) = \max_{\omega \in \Omega}(\max_{\pi \in \Pi} b(\omega, \pi) - \min_{\pi \in \Pi} b(\omega, \pi)).$$

For a given graph G the function $GAP(G)$ is a measure of the maximum bias of an electoral outcome in terms of number of seats in single member majority districts.

PROPOSITION 4 $GAP(G) \leq 2W(G) - p = 2\lfloor \frac{n}{s+1} \rfloor - p.$

Proof. Since $b(\omega, \pi) + r(\omega, \pi) = p$, we have

$$GAP(G) = \max_{\omega \in \Omega}(\max_{\pi \in \Pi} b(\omega, \pi) + \max_{\pi \in \Pi} r(\omega, \pi)) - p \leq \qquad (1)$$

$$\max_{\omega \in \Omega} \max_{\pi \in \Pi} b(\omega, \pi) + \max_{\omega \in \Omega} \max_{\pi \in \Pi} r(\omega, \pi) - p = 2W(G) - p.$$

∎

For a given p-equipartitionable graph G we are interested in finding, if it exists, a bicoloring $\omega^* \in \Omega$ such that there are both a blue extremal partition and a red extremal one, both w.r.t. ω^*. If such a bicoloring exists, we will say that G is *two-faced* and there exist two partitions $\pi_b, \pi_r \in \Pi$ such that:

$$b(\omega^*, \pi_b) = r(\omega^*, \pi_r) = W(G) = \lfloor n/(s+1) \rfloor.$$

COROLLARY 5 *We have*

$$GAP(G) = 2W(G) - p \qquad (2)$$

if and only if G is two-faced.

Proof. Follows from (1). ∎

Two-faced graphs are those for which gerrymandering exhibits its worst case bias. There is an absolute threshold for the largest number of seats that a party can obtain when the vote outcome is balanced. In two-faced graphs, for a suitable balanced vote, both parties can achieve this threshold by artful gerrymandering.

3.   Theoretical Results on Grid Graphs

The main result of this section is that, under the above assumptions on n, s, and p, any grid graph with an even number of vertices is two-faced.

Let G be a grid graph with M rows and N columns, and $n = MN$. Since we assume that n is even, at least one between M or N must be even. In the following we assume, without loss of generality, that M is even.

Even grids feature one simple property which is crucial for the development of the results to follow: they are hamiltonian (see Figure 2). On the one hand, this property implies that even grids are p-equipartitionable, since obviously a cycle of length $n = ps$ can always be partitioned into p paths of length s (remember that p-equipartitionability is NP-complete for general graphs). On the other hand, in an even cycle there are only s partitions into subpaths of the same size s. Each of them results from cutting p equidistant edges of the cycle, and thus it can be easily obtained from the others by a suitable rotation of the cuts along the cycle. If one can show that there exists one such partition satisfying certain properties, then this is sufficient to establish the existence in an even grid of a district map satisfying the same properties. This tool will be often exploited in our constructions.

We start from the case $p = s + 1$, where a blue extremal partition has exactly s edgy districts and one sweeping district. In fact, by Corollary 2 with $q = 0$ and $r = s + 1$, the upper bound on the number of blue districts is s. These districts must be edgy since the number of blue vertices in G is $s(s + 1)/2$. It follows that the remaining district is red sweeping. We will show how to construct such an extremal partition on a hamiltonian cycle H of G. We suppose that the vertices of H are consecutively numbered from 1 to n along the cycle (traversed clockwise).

A *boa* is a path with $(s + 1)(s - 1)/2$ vertices that can be partitioned into $(s+1)/2$ components having $(s-1)/2$ consecutive blue vertices and $(s-1)/2$

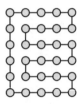

Fig. 2.   Hamiltonian cycle in a grid graph with an even number of rows.

Fig. 3.   Examples of boas.

consecutive red vertices each. Boas have the following nice property: if one cuts the s-th, the $2s$-th, ... , the $((s-1)s/2)$-th edge from left to right, one obtains $(s-1)/2$ red edgy districts and the remaining $(s-1)/2$ nodes are blue; a symmetrical property holds when one interchanges the two colors "red" and "blue", as well as "right" and "left".

In Figure 3 the boas for $s = 5$ and $s = 7$ are shown. Here, as in all black and white figures in the sequel, blue vertices are displayed in white and red vertices in black.

In Figure 4 we consider the case $s = 5$ and we show how to use two boas in order to find a bicoloring of H for which there are both a blue extremal partition and a red extremal one. One obtains such bicoloring by splitting H into four consecutive subpaths that are colored in the following way:

- the first subpath P_1 extends from vertex 1 to vertex $(s+1)/2$ and all its vertices are red;

- the second subpath P_2 is a boa starting from the red vertex $(s+1)/2+1$ and ending at the blue vertex $s(s+1)/2$;

- the third subpath P_3 extends from vertex $s(s+1)/2 + 1$ to vertex $(s+1)(s+1)/2$ and all its vertices are blue;

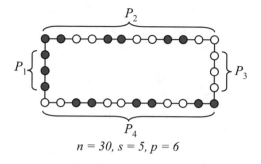

$n = 30, s = 5, p = 6$

Fig. 4.   Bicoloring for the case $p = s + 1$.

- the fourth subpath P_4 is a boa starting from the red vertex $(s + 1)(s + 1)/2 + 1$ and ending at the blue vertex $s(s + 1)$.

It is easy to verify that the number of blue vertices is equal to the number of red ones. Since H is a cycle, one can obtain an arbitrary partition into p connected components by cutting p edges. In Figure 5 the two extremal partitions are shown for the case $s = 5$. If the cut edges are $(s, s+1), (2s, 2s+1), ..., (s^2, s^2+1), ((s+1)s, 1)$ the district containing vertices from 1 to s is red sweeping and all the other ones are blue edgy (Figure 5 (a)). Thus the partition is blue extremal. By shifting each cut to its next edge (clockwise) $(s + 1)/2$ times, we obtain a blue sweeping district from vertex $s(s + 1)/2 + 1$ to vertex $s(s + 1)/2 + s$ and all the other districts are red edgy. So the partition is red extremal (Figure 5 (b)).

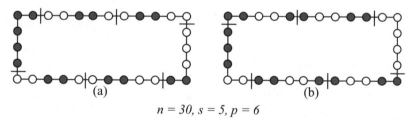

$n = 30, s = 5, p = 6$

Fig. 5.   Partitions for the case $p = s + 1$.

Let us consider now the case $p < s+1$. Since p is even and positive we can suppose $p = (s+1) - 2k$ for some k such that $1 \le k \le (s-1)/2$. As shown in Figure 6 for the case $s = 5$ and $k = 1$, starting from the bicoloring of the case $p = s + 1$ we delete from the subpath P_2 the last ks vertices and from the subpath P_4 the first ks vertices. We obtain a cycle with $s(s+1) - 2ks$ vertices where the number of blue vertices is equal to the number of red ones. If one cuts the edges as above, starting from $(s, s+1)$, the district containing vertices from 1 to s is red sweeping and all the other ones are blue edgy except the one containing the subpath P_3, which is not edgy because it contains $(s+1)/2 + k$ blue vertices and $(s - 1)/2 - k$ red vertices. The obtained partition is blue extremal. By shifting the cuts as for the case $p = s + 1$, the resulting partition is red extremal. In fact, in the district containing the subpath P_3, the blue party wins since there are $s - k$ blue vertices and k red vertices, while all the other districts are red edgy.

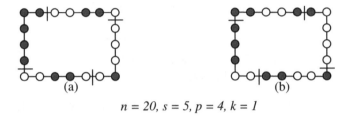

$n = 20,\ s = 5,\ p = 4,\ k = 1$

Fig. 6.   Bicoloring and Partitions for the case $p < s + 1$.

Finally suppose that $p > s + 1$.

PROPOSITION 6 *Under the above assumptions on M, N, p and s, G can be decomposed into p grid subgraphs having s vertices each.*

Proof. Since $MN = ps$ there exist four natural numbers M_1, M_2, N_1 and N_2 such that:

$$M = M_1 M_2,\ N = N_1 N_2,\ M_1 N_1 = s,\ M_2 N_2 = p.$$

As shown in Figure 7 (a), by partitioning the columns of G into N_2 components having N_1 columns each and the rows of G into M_2 components having M_1 columns each, one can decompose G into p grid subgraphs having M_1 rows and N_1 columns each. Notice that, since s is odd, also M_1 and N_1 are odd; hence, since M is even, also M_2 is even. ∎

As in Corollary 2, we suppose that $p = q(s + 1) + r$, with $q \geq 1$ and $1 \leq r \leq s + 1$. Notice that, since $s + 1$ and p are even, also r must be even.

We represent the decomposition given in Proposition 6 by a grid graph $\overline{G}$, with M_2 rows and N_2 columns, whose vertices V_k, $k = 1, ..., p$, correspond to the grid subgraphs and there is an edge connecting the vertices V_k and V_j if

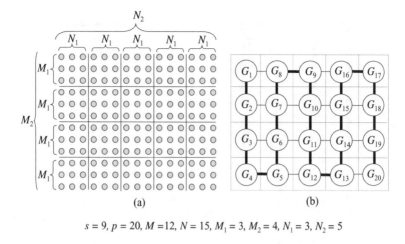

$s = 9,\ p = 20,\ M = 12,\ N = 15,\ M_1 = 3,\ M_2 = 4,\ N_1 = 3,\ N_2 = 5$

Fig. 7.   Decomposition of G into p grid subgraphs. The hamiltonian path is marked bold.

some vertex of the grid corresponding to V_k is adjacent to some vertex of the grid corresponding to V_j (see Figure 7 (b)). Let us consider the hamiltonian path $\overline{P} = (V_1, V_2, ..., V_p)$ of $\overline{G}$ and partition it into q subpaths having $s + 1$ vertices each and one subpath having r vertices. Let P_j be the j-th subpath of $\overline{P}$.

LEMMA 7 *For each* $j = 1, ..., q + 1$, *and for each column* c *of* $\overline{G}$, *the number of vertices of* P_j *in column* c *is even.*

Proof. The proof is based on the fact that the number of rows of $\overline{G}$, M_2, and the number of vertices in each subpath P_j, $s + 1$ or r, are even. Let c_1 be the smallest numbered column whose intersection with some of the subpaths P_j is odd. Then c_1 must intersect in an odd number of nodes an even positive number of subpaths P_j. But then the smallest numbered such subpath, by the minimality assumption on c_1, would contain an odd number of nodes, a contradiction. ∎

As shown in Figure 8, the subpaths P_j, $j = 1, ..., q + 1$, define in G a decomposition into $q + 1$ connected subgraphs $H_1, ..., H_{q+1}$.

PROPOSITION 8 *For each* $j = 1, ..., q + 1$, H_j *is hamiltonian.*

Proof. As shown in Figure 8, each H_j can be decomposed into at most three grid subgraphs which, by Lemma 7, have an even number of rows. Hence it is possible to find a hamiltonian cycle of H_j as in the graph of Figure 9. ∎

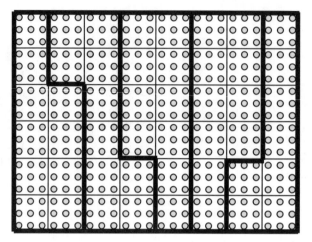

$s = 9, p = 48, M = 18, N = 24, M_1 = 3, M_2 = 6 \, N_1 = 3, N_2 = 8$

Fig. 8.   Decomposition of G into $q + 1$ hamiltonian subgraphs.

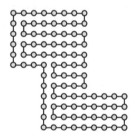

Fig. 9.   Hamiltonian cycle in a H_j subgraph of G.

Since H_j, $j = 1, ..., q + 1$ is hamiltonian, then, as shown before, it is two-faced and so it is possible to find a bicoloring such that there exist a blue extremal partition and a red extremal one. By using the blue extremal partitions of the subgraphs H_j, one can obtain a partition of G having $qs + r - 1$ blue districts. In fact, by Corollary 2, in each of the q subgraphs having $s(s + 1)$ vertices, there are s blue districts and in the subgraph having rs vertices there are $r - 1$ blue districts. But, again by Corollary 2, $qs + r - 1$ is an upper bound on $W(G)$, hence the partition of G is blue extremal. The same arguments can be used for obtaining a red extremal partition. Then G is two-faced.

By the constructions shown for the cases $p = s + 1$ and $p < s + 1$ and the decomposition found for the case $p > s + 1$, the following theorem holds.

THEOREM 9 *Under the above assumptions on p and s, any grid graph with ps vertices is two-faced.*

COROLLARY 10 *If $G(s+1, s)$ is a grid graph with $s+1$ rows and s columns, then*

$$\lim_{odd\ s\to\infty} \frac{GAP(G(s+1,s))}{s+1} = 1.$$

Proof. After Theorems 5 and 9, one has

$$\frac{GAP(G(s+1,s))}{s+1} = \frac{2W(G(s+1,s)) - s - 1}{s+1} = \frac{2s - s - 1}{s+1} = \frac{s-1}{s+1}.$$

When s odd $\to \infty$, the thesis follows. ∎

Corollary 10 is stunning: it means that, for certain infinite families of grids, as the number and size of the districts grow, vicious gerrymandering can make the percentages of blue districts and red ones both arbitrarily close to 1 even under the assumptions that the vote outcome is the same and that the blue party and the red one get the same total number of votes.

In conclusion, we have shown that for all even grids one can construct Dixon-Plischke-like examples where gerrymandering can heavily reverse the electoral result in terms of Parliament seats.

Our final result shows that for some highly symmetric colorings, on the one hand, there are blue and red extremal district designs; on the other hand, the most compact design, namely, the partition of the grid into square subgrids, yields the same number of blue and red districts.

To address the question we introduce the notion of *skew-symmetric* coloring.

Let φ be the mapping of the grid onto itself that maps node (i, j) into $(M + 1 - i, N + 1 - j)$. Notice that φ is the product of two reflections, the first one around the y-axis, the second one around the x-axis. Since M is even, φ fixes no point of G. A coloring $\omega \in \Omega$ is *skew-symmetric* if (i, j) and $\varphi(i, j)$ have opposite colors.

If a grid is skew-symmetrically colored, then $\varphi(G)$ is isomorphic to G, the colors of its vertices being interchanged (in fact φ is an automorphism of the grid). In other words, up to the labels of the vertices, the effect of φ on G reduces to switching the colors of its vertices.

THEOREM 11 *Let G be an M × N grid having ps vertices with $p \leq s+1$ and p even. One can always find a blue- and a red- extremal partition with respect to some skew-symmetric bicoloring of G.*

Proof. (Sketch). We can divide the grid into two equally sized parts, say L and R, of $\frac{ps}{2}$ vertices each, in such a way that: (i) $(i,j) \in L$ if and only if $\varphi(i,j) \in R$; (ii) both L and R induce subgraphs containing hamiltonian paths.

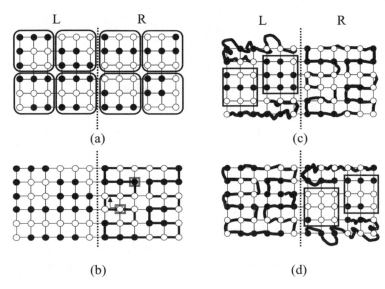

Fig. 10. **(a):** The most compact and equitable partition of a 6×12 skew-symmetrically colored grid. **(b):** The hamiltonian cycle from which the two extremal partitions in (c) and (d) are generated. Starting from the framed blue (white) vertex, and cutting the 9th, 18th, 27th and 36th edges of the cycle (clockwise) the right hand side of the partition in (c) is generated (the left hand side of the partition in (d) can be obtained by symmetry). Similarly, the right hand side of the partition in (d) (and, by symmetry, the left hand side of the partition in (c)) is generated by starting from the framed red (black) vertex. **(c),(d):** Red and blue extremal partitions.

Let us consider the subgraph G_R induced by R. We can define a coloring of G_R and two connected partitions π'_R and π''_R into $p/2$ components such that: π'_R is a partition all whose districts are red edgy, π''_R is a partition all whose districts but one are blue edgy, the exceptional one being red (see Figure 10). Using φ we extend the coloring of G_R to the entire grid. By construction this coloring is skew-symmetric. Moreover, if C is any component of either π'_R or π''_R, $\varphi(C)$ is a connected component of G_L (the graph induced by L), isomorphic to C but with colors interchanged. It follows that if π'_L and π''_L are the partitions of G_L corresponding via φ to π'_R and π''_R, respectively, then $\pi'_R \cup \pi''_L$ and $\pi''_R \cup \pi'_L$ are extremal partitions of G. ∎

However, skew-symmetric colorings give rise not only to maximally biased designs, but also to minimally biased compact designs (see Figure 10).

THEOREM 12 *Let G be a skew-symmetrically colored $M \times N$ grid. Suppose that G can be divided into squares of sides of length $\sqrt{s}$ and let π be the partition formed by such squares. Then, in π, the number of red district equals the number of blue districts* .

Theorem 11 shows that even highly symmetrical vote outcomes can be manipulated in a partisan way. Nevertheless, in view of Theorem 12, compactness can be considered (at least within the frame of our idealized model) as an effective remedy against gerrymandering.

4.   Experimental Results on Real-Life Test Problems

In this section we provide a multiobjective graph partitioning model for political districting and we study gerrymandering from an experimental point of view on real-life data. Starting from the graph-theoretic model described in Section 2, here we relax some of the previous assumptions in order to adhere to reality as much as possible. In both models the territory is represented by a graph and one looks for a connected partition of the graph in order to enforce the integrity and contiguity requirements. In the previous sections the underlying graph was assumed to be a rectangular grid, while here it may be, more generally, an arbitrary planar graph. In the former model nodes were unweighted and a vote outcome was but a node bicoloring; here nodes are weighted both by their populations and their votes. Whereas the stylized previous model is more amenable to theoretical investigation, the one we shall study in this section is more flexible and offers a more accurate description of real-life political districting. It is no coincidence that variants of it have been considered by several Authors (Bussamra et al., 1996, Garfinkel and Nemhauser, 1970, Merhotra et al., 1998, Nygreen, 1988, Ricca and Simeone, 2005). In spite of their differences, both models lead, in different ways, to the same conclusions: gerrymandering can drastically reverse the final outcome of an election, and compactness does provide an effective protection. Thus the experimental results of the present section corroborate and validate the theoretical results obtained so far.

Remember that our aim is to investigate both how bad gerrymandering can be and, simultaneously, to determine if there exist effective weapons against it, such as compactness or other districting criteria, which can be adopted in order to avoid this practice. Thus, in our experiments we undertake a multicriteria approach and develop an optimization model in which different objective functions - measuring different criteria - are considered one at a time.

4.1   The Model and the Database

As before, n denotes the total number of territorial units in the territory, $n = |V|$, and p, $1 \leq p \leq n$, is a positive integer denoting the number of districts. Let p_i, $\forall i \in V$, be positive integral node-weights, representing territorial unit populations and d_{ij}, $\forall i, j \in V$, be positive real distances defined for each unit pair (i, j). For each territorial unit, the list of all those administrative areas (regions, provinces,...) that contain the unit is known. Finally, with reference to political elections in Italy, for each territorial unit we introduce two positive integral node weights, vo_i and vp_i, $\forall i \in V$, representing the number of votes obtained in unit i by the Olive Tree and by the Pole of Liberties, respectively[2].

The general graph partitioning problem can be formulated as follows:
Given a graph G, partition its set of nodes into p subsets (districts) such that the subgraph induced by each subset is connected and a given function of the partition is optimized.

In the sequel, we use the term "district design" as a synonym of "connected partition into p components". Actually, we are no longer imposing the further restriction that the districts be equally sized since in real-life cases this requirement is too strict and we can only try to get close to the ideal case as much as possible by optimizing a suitable objective function.

In our experiments we used data of three Italian Regions, namely, Piedmont, Latium and Abruzzi, whose townships are taken to be the territorial units. The weights p_i associated to territorial units correspond to the Italian population from 1991 Census, and we considered the real road distances between pairs of territorial units. In this application we considered the Italian (majoritarian) vote distribution of Political Elections of 1996.

4.2   Districting Criteria and Local Search Algorithm

In our real life model we considered several of the most commonly adopted districting criteria discussed in Section 1. In particular, integrity and contiguity are automatically guaranteed by the graph-theoretic model in which each elementary territorial unit is represented by a vertex of the graph. The remaining criteria of population equality (PE), compactness (C) and conformity to admin-

---

[2]In this application we consider the Italian (majoritarian) vote distribution of Political Elections of 1996. The Olive Tree and Pole of Liberties parties were the center-right and center-left coalitions, respectively, which were in competition at that time.

istrative boundaries (AC) are measured by proper indicators to be optimized. To this purpose, we adopted some indicators already used in similar applications (Grilli di Cortona et al., 1999). Actually, they measure non-population equality, non-compactness and non-administrative conformity, therefore they must be minimized. Firstly, an ideal situation of perfect population equality, perfect compactness and perfect administrative conformity is defined and a proper index is chosen so that its value is equal to 0 when the ideal situation is met, while in the other cases it provides a measure of the corresponding error. These indexes are generally normalized in order to be independent of scaling factors. Therefore, they can be read as percentages.

Let $C_1, C_2, \ldots, C_p \subset V$ be the subsets of nodes of the p districts of a given district design. Let $P_k = \sum_{i \in C_k} p_i$, $k = 1, 2, \ldots, p$, be the population of district C_k. Then, the population equality index for the district design is given by

$$PE = \frac{\sum_k |P_k - \bar{P}|}{p\bar{P}} \tag{3}$$

where $\bar{P} = \frac{\sum_k P_k}{p}$ is the average district population. This is the average deviation of the population of each district from $\bar{P}$, divided by the normalization factor $\bar{P}$.

On the basis of the distances d_{ij}, for each pair of vertices $i, j \in V$, we define a global compactness index given by the sum of compactness indices computed over each district separately. For a given district C it can be briefly described as follows. Let d_{ij} be the distance between unit i and unit j. For each unit compute its eccentricity

$$d(i) = \max_{j \in C} d_{ij}$$

and set

$$\delta = d(s) = \min_{i \in C} d(i)$$

By definition, s is the center of district C and the compactness in district C is measured by:

$$C = \frac{\sum_{i \in C} p_i}{\sum_{j \in D} p_j} \tag{4}$$

where $D = \{j \in V : d_{js} \leq \delta\}$.

The compactness index (4) is a measure of the deviation of the districts from the ideal situation in which they all have a regular, "round" shape.

The administrative conformity index adopted in this application is defined on the basis of the discrepancies between the already existing administrative district maps (of different type) and the electoral district design. For a given district and a given type of administrative boundaries it is basically computed as a measure of those units which produce discrepancies. The global index, which varies between 0 and 1, is obtained by averaging over all types of administrative boundaries and over all the electoral districts. A detailed description of the index is reported in (Grilli di Cortona et al., 1999).

We considered these three indexes as objective functions in our optimization model, both separately and combined together into a single objective function given by a convex combination of them.

Notice that the population equality index defined for the real-life application can be considered as the counterpart of the principle of equal size districts stated in the combinatorial model of Section 2. In our graph-theoretic model a vertex corresponds to an elementary unit of the territory. In general - as in our case - territorial units are given by townships and it is not guaranteed that they have the same size. Thus, the requirement of Section 2 which forces each district to have exactly the same number of units as each other here does not work. Actually, the *one-man-one-vote* principle addressed by that assumption here must be necessarily pursued through population equality, regarded not as a hard constraint, but as a criterion to be fulfilled as much as possible.

On the other hand, the idea of compact districts sketched in Section 3 (see Figure 10) perfectly matches the principle embedded in our compactness index (4).

The additional administrative conformity index was considered in our experiments since it is generally included among the commonly and widely accepted political districting criteria. The experimental results related to it add some more information to our knowledge and can be useful for evaluating the actual relevance of this criterion in a districting procedure.

Since we are interested in studying how far gerrymandering can be pushed, we must also consider partisan criteria. Here we are obliged to adhere to reality and, in our case, we refer to the real vote distribution of Italian political elections of 1996. With respect to this vote, for any given district design, we are able to compute how many seats are assigned to the Pole and to the Olive party, respectively. The idea is that both Pole and Olive would like to win the election. To this purpose, if they each had the opportunity of designing their own political districts, they would try to find the district design that makes them win as many seats as possible (gerrymandering). Actually, for a given district design, the number of seats assigned to a party can be considered as a measure of the *utility* of the district design for that party. For a given district, let ρ be the ratio between the number of votes for the Pole and those for the

Olive. Then, the utility of this single district for the Pole can be measured through the following step function:

$$h(\rho) = \begin{cases} 0, & \text{if } \rho < 1 \\ 1, & \text{if } \rho \geq 1, \end{cases} \tag{5}$$

and the sum of such district utilities for the Pole over all the districts provides a partisan index for the Pole. Similarly, we can compute a partisan index for the Olive party. These two indexes can be adopted as objective functions in our experimental analysis when we study how far the Pole and the Olive party can manipulate the district design, respectively.

In our experiments we used the Old Bachelor Acceptance metaheuristic (Hu et al., 1995) in order to find solutions that minimize the six different objectives. This metaheuristic has shown to perform well when applied to territorial political districting problems. For details, see (Ricca and Simeone, 2005).

We notice here that local search techniques are particularly suitable also for the design of partisan districts. Actually, starting from an initial district design, they work by performing small perturbations of the current solution. At each step a node belonging to the boundary of a district migrates towards an adjacent district. Thus, two consecutive district designs differ just for one node in only two districts and it is hard to distinguish between them. Migration by migration, it is possible to obtain a district design which favors a given party (its utility is maximized) and such that the initial given district design is modified as little as possible. However, when applying local search techniques, (5) is not sufficiently sensitive to the migration of a vertex from one district to another. This explains why we chose to replace the step function (5) by a smoother objective function. For a given party, say the Pole, in each district we compute the following district-utility *logistic* function for that party

$$g(\rho) = \frac{c}{1 + exp(b\,(1 - \rho))},$$

where c and b are suitably chosen in order to get the desired shape of the utility function. The idea is that the district-utility increases rapidly when ρ is near 1 (see Figure 11).

The aim of our experimental study is twofold. On the one hand, we want to test if our four objective functions, given by PE, C and AC, and their convex combination, are good weapons against gerrymandering. On the other hand, starting from a given district design we try to manipulate it as much as possible in order to maximize the objective function given by the utilities of the Pole and the Olive party, respectively. The underlying idea is that gerrymandering can be investigated experimentally in order to identify worst case configurations and the corresponding upper bounds over the maximum number of seats that a party can get. From our previous experimental works, we already know

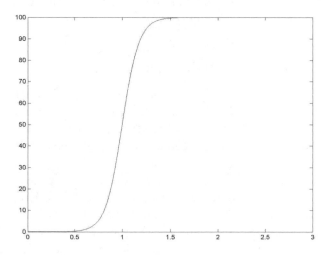

Fig. 11.   District-utility logistic function for $c = 100$ and $b = 11, 51$.

the good performance of PE, C and AC as objective functions in non-partisan districting problems. However, in this paper the results referring to the manipulation of the districts are new. Moreover, as we will see in the next section, our experimental results show that our neutral objectives can be used as an alarm signal for gerrymandering, since they tend to deteriorate when gerrymandering is practiced.

It is clear that our experimental results cannot be compared to the exact bounds given in Section 2. However, we believe that these results are of interest on their own because they represent the real-life counterpart of our theoretical results of the previous sections. Therefore, such results are not a mere mathematical curiosity, but they capture the gist of the real threat posed by gerrymandering. As we will see in Section 4.3, there are regions in which, for suitable district designs, the Pole or the Olive party gets the total number of seats.

4.3   Experimental Plan and Results

Table 1 shows the main characteristics of the graphs representing the territories of three Italian regions considered in our experimental plan.

As before, here PE means "Population Equality", C means "Compactness" and AC means "Administrative Conformity", while MT refers to the "Mixed Target" which is defined as the following convex combination of PE, C and

Table 1.   Graphs of the Italian Regions

Region	N. of Nodes	N. of Edges	Density	N. of Districts
Piedmont	1208	3527	2.92	28
Latium	374	1006	2.69	19
Abruzzi	305	847	2.78	11

AC:
$$0.5PE + 0.3C + 0.2AC.$$

For each region we performed six different runs of Old Bachelor Acceptance metaheuristic with PE, C, AC, MT and the two utility functions as objectives, respectively. Following (Ricca and Simeone, 2005), we implemented a randomized version of this metaheuristic, that is, starting from an initial solution, at each iteration Old Bachelor Acceptance chooses a random solution in the neighborhood of the current one. Notice that randomization is a useful tool for the diversification of the search: it is used to avoid cycling and explore a large amount of different solutions. When the objectives are PE, C, or AC, the initial solution is generated randomly. After a spanning tree T of G is randomly generated, $p - 1$ randomly chosen edges of T are cut in order to get p subtrees whose node-sets correspond to the p initial districts. For the MT criterion we preferred to start from the district map generated by the ADEN heuristic in (Grilli di Cortona et al., 1999).

The optimal solutions found in the previous four runs were adopted as possible initial solutions for the case in which the objective is to maximize the utility of a given party. The idea was that starting from an already optimized set of districts could make it more difficult to manipulate the given district design in favor of one of the two parties. However, also the Institutional district design of the Italian Political Elections of 1996 was considered as possible initial solution. Among the results obtained w.r.t. these 5 different initial district designs, we selected the worst observed case.

Tables 2-4 show our experimental results on the three different graphs. The last row of Tables 2-4 refers to the values of the six objectives computed for the Institutional district design adopted in Italy for the Political Elections of 1996. This row was included in order to favor the comparison between our - neutral and partisan - district designs and the one that was actually adopted in 1996.

On the basis of our experiments, we can state the following conclusions:

1. Given a vote distribution, gerrymandering is able to dramatically reverse the electoral outcome (see, the fifth and the sixth row of each table).

Table 2. Piedmont

District Design	PE	C	AC	MT	Pole seats	Olive seats
Min PE	**0.075**	0.911	0.577	0.426	10	18
Min C	0.771	**0.531**	0.347	0.614	11	17
Min AC	0.940	0.643	**0.113**	0.686	12	16
Min MT	0.094	0.762	0.288	**0.334**	11	17
Max Pole	1.052	0.777	0.454	0.850	**21**	7
Max Olive	1.364	0.593	0.263	0.913	3	**25**
Institutional	0.105	0.859	0.143	0.339	11	17

Table 3. Latium

District Design	PE	C	AC	MT	Pole seats	Olive seats
Min PE	**0.046**	0.778	0.523	0.361	13	6
Min C	1.226	**0.166**	0.143	0.692	12	7
Min AC	1.072	0.620	**0.050**	0.732	13	6
Min MT	0.050	0.502	0.270	**0.230**	10	9
Max Pole	1.512	0.321	0.061	0.864	**19**	0
Max Olive	1.299	0.277	0.131	0.759	3	**16**
Institutional	0.060	0.683	0.202	0.275	10	9

Table 4. Abruzzi

District Design	PE	C	AC	MT	Pole seats	Olive seats
Min PE	**0.040**	0.744	0.508	0.345	4	7
Min C	0.668	**0.390**	0.288	0.508	4	7
Min AC	0.894	0.539	**0.056**	0.620	4	7
Min MT	0.113	0.442	0.263	**0.242**	4	7
Max Pole	1.217	0.425	0.320	0.800	**10**	1
Max Olive	1.129	0.473	0.328	0.772	1	**10**
Institutional	0.078	0.633	0.215	0.272	5	6

2. The districting bias produced by gerrymandering algorithms implies the deterioration of the values of all the traditional PD criteria.

3. It turns out that there is a substantial stability of the number of seats attributed to the Pole and to the Olive when the criteria of Population Equality, Compactness, Administrative Conformity and the Mixed one are optimized.

4. Compactness is a good shield against the practice of gerrymandering. On the other hand, in view of 3, and since gerrymandering deteriorates *all* the districting criteria, satisfying the other criteria helps in preventing gerrymandering. This is why the use of more than one traditional PD criterion is generally recommended.

References

M. Balinski, Le suffrage universel inachevé, Belin, Paris, 2004.

N. M. Bussamra, P. M. França, N. G. Sosa, Legislative districting by heuristic methods, AIRO 1996 Procedings, pp. 640-641.

R. J. Dixon, E. Plischke, American Government: Basic Documents and Materials, New York, Van Nostrand, 1950.

M.E. Dyer, A.M. Frieze, On the complexity of partitioning graphs into connected subgraphs, Discrete Applied Mathematics, 10 (1985), pp. 139-153.

R. S. Garfinkel, G. L. Nemhauser, Optimal political districting by implicit enumeration techniques, Management Science, 16 (1970), pp. 495-508.

P. Grilli di Cortona, C. Manzi, A. Pennisi, F. Ricca, B. Simeone, Evaluation and Optimization of Electoral Systems. SIAM Monographs in Discrete Mathematics, SIAM, Philadelphia, 1999.

T. C. Hu, A. B. Kahng, and C. W. A. Tsao, Old Bachelor Acceptance: a new class of non-monotone threshold accepting methods, ORSA Journal on Computing, 7 (1995), pp. 417-425.

A. Merhotra, E. L. Johnson, and G. L. Nemhauser, An optimization based heuristic for political districting, Mangement Science, 44 (1998), pp.1100-1114.

B. Nygreen, European Assembly constituencies for Wales. Comparing of methods for solving a political districting problem, Mathematical Programming, 42 (1988), pp. 159-169.

F. Ricca, B. Simeone, Local Search Algorithms for Political Districting, submitted to European Journal of Operational Research.

Apportionment: Uni- and Bi-Dimensional

Michel Balinski
Laboratoire d'Économétrie, École Polytechnique

Abstract   This paper characterizes divisor methods for vector and matrix apportion problems with very simple properties. For the vector problem—a vector gives the votes of parties or the populations of states, a single number the size of the house—they are shown to be the only methods that are *coherent with* the definition of the corresponding divisor method when applied to only two states or parties. For the matrix problem—rows correspond to districts, columns to parties, entries to votes for party-lists, and the number of seats due to each row (or district) and each column (or party) is known—one extra property is necessary. The method must be *proportional*: it must give identical answers to a problem obtained by re-scaling any rows and/or any columns of the matrix of votes.

Keywords: Apportionment, divisor method, coherence, biproportional apportionment, rounding, justified rounding.

1.   Introduction

"Bi-dimensional" (or "matrix") apportionment is now a recognized system for designating winners in an election system. It is the law of the land in the Swiss canton and the city of Zürich (Pukelsheim and Schuhmacher, 2004), and it may well become so in the Faroe Islands (Zachariassen and Zachariassen, 2005). Developed, justified, explained and applied in a series of papers and a book (Balinski and Demange, 1989a,b; Balinski and Rachev, 1997; Balinski and Ramírez, 1997,1999a; Balinski, 2002, 2004) it may also be viewed as a simple and direct extension of the more familiar "uni-dimensional" (or "vector") apportionment problem. That is what this paper aims to do.

2.   Vector Apportionment: a Primer

A *vector (or uni-dimensional) apportionment problem* is a pair (v, h), where $v = (v_i) > 0$ for $i = 1, \ldots, m$ are the populations of m regions (or the votes of m parties) and h is the number of seats in an assembly to be distributed "proportionally" among them. An *apportionment* is a vector $a = (a_1, \ldots, a_m)$, where $a_i \geq 0$ is integer valued and $\sum_i a_i = h$. Vector apportionment is the classical problem of allocating seats to regions or states when v is the vector of their populations, or of allocating seats to political parties when v is the vector of their votes: by what method should a solution be chosen from among the many possible apportionments?

In general, a *(vector) method of apportionment* Φ selects a nonempty subset of apportionments $\Phi(v, h)$ for any problem (v, h).

A *divisor criterion* is any real valued function d on the nonnegative integers $k \geq 0$ that satisfies $k \leq d(k) \leq k + 1$ and for which there are no two integers $p > 0$ and $q \geq 0$ where $d(p) = p$ and $d(q) = q+1$. In effect, a divisor criterion is simply a point on each closed interval $[k, k + 1]$ for $k \geq 0$ and integer, with the stipulation that if in some interval the point is at the lower (the upper) end then in no other interval can it be at the upper (the lower) end. Suppose that a real number x is in the interval $[a, a + 1]$, a an integer. Then a *d-rounding* $[x]_d$ of $x > 0$ is a if $x < d(a)$ and $a + 1$ if $x > d(a)$; if $x = d(a)$ then $[x]_d$ is either a or $a + 1$ (so in fact $[x]_d$ is a set that is usually single valued). A d-rounding of 0 is always 0: $[0]_d = 0$. The $d(a)$ are thresholds in the intervals $[a, a + 1]$: below the threshold x is rounded-down to a, above it is rounded-up to $a + 1$, at the threshold it is either rounded-up or -down.

A *divisor method based on d* is the set of apportionments

$$\Phi_d(v, h) = \left\{ a = (a_i) : a_i = [\lambda v_i]_d \text{ for } \lambda \text{ chosen so that } \sum_i a_i = h \right\}. \quad (1)$$

If, contrary to the definition, $d(p) = p, d(q) = q + 1$ for some integers $p > 0, q \geq 0$, then $(p-1, q+1) \in \Phi_d((p, q), p+q)$, showing that although a perfect apportionment exists it may not be chosen, and explaining the exclusion. Note also that $d(0) = 0$ implies $[\lambda v_i]_d \geq 1$ for every $\lambda > 0$ and $v_i > 0$.

If $a \in \Phi_d$ then $d(a_i - 1) \leq \lambda v_i \leq d(a_i)$ for all i, implying

$$\Phi_d(v, h) = \left\{ a = (a_i) : \min_{a_i > 0} \frac{v_i}{d(a_i - 1)} \geq \max_{a_j \geq 0} \frac{v_j}{d(a_j)}, \sum_i a_i = h \right\}, \quad (2)$$

where $v_j/0 = \infty$ and $d(-1) = 0$. Consequently, a divisor method may also be described recursively as follows. $\Phi_d(v, 0) = 0$ and suppose $a \in \Phi_d(v, h)$. Then

$$\bar{a} \in \Phi_d(v, h + 1) \text{ where } \bar{a} = a, \text{ except } \bar{a}_l = a_l + 1 \text{ for } \frac{v_l}{d(a_l)} = \max_i \frac{v_i}{d(a_i)}. \quad (3)$$

This description implies that $a \in \Phi_d(v, h)$ if a solves

$$\max_a \min_i \frac{v_i}{d(a_i - 1)} \text{ when } \sum_i a_i = h \text{ and } a_i \geq 0 \text{ integer.} \quad (4)$$

From the recursive definition it also follows that an apportionment a of the divisor method Φ_d is a solution to

$$\max_a \sum_i \sum_{k=0}^{a_i-1} \frac{v_i}{d(k)} \text{ when } \sum_i a_i = h \text{ and } a_i \geq 0 \text{ integer},\qquad(5)$$

(assuming $h \geq m$ if $d(0) = 0$). This is easy to see because the solution is "greedy": at each allocation of an "extra" seat give to the integer variable k the value that maximizes $v_i/d(k)$ over i in conformity with the recursive procedure. Precisely the same argument shows that an apportionment a of the divisor method Φ_d is also a solution to

$$\max_a \prod_i \prod_{k=0}^{a_i-1} \frac{v_i}{d(k)} \text{ when } \sum_i a_i = h \text{ and } a_i \geq 0 \text{ integer}.\qquad(6)$$

There are many other "objective functions" that are optimized by the apportionments of one or another of the divisor methods.

The *parametric (divisor) method* Φ_δ based on δ, for $0 \leq \delta \leq 1$, is the divisor method Φ_d based on d where $d(k) = k + \delta$ for all integer $k \geq 0$. Adams's method is the parametric method based on $\delta = 0$; Condorcet's method is the parametric method based on $\delta = \frac{2}{5}$; Webster's or Sainte-Laguë's is based on $\delta = \frac{1}{2}$; and Jefferson's or D'Hondt's is based on $\delta = 1$.

Letting $\bar{v} = \sum v_i$, an apportionment a of the parametric method Φ_δ is a solution to (see Balinski and Ramírez (1999b)):

$$\min_a \sum_i v_i \left(\frac{a_i + \delta - \frac{1}{2}}{v_i} - \frac{h}{\bar{v}} \right)^2 \text{ when } \sum_i a_i = h \text{ and } a_i \geq 0 \text{ integer}.\quad(7)$$

Notice that solutions to the optimization problems (4), (5) and (6) do not change when v is replaced by λv for any $\lambda > 0$.

There are an infinite number of divisor methods for vector problems, and they can yield very different apportionments. They have been characterized by a set of properties so desirable in the context of apportionment that it is fair to say they are the *only* acceptable methods (Balinski and Young, 1982). The most important of these properties—coherence—stems from a very simple idea.

Suppose that the h seats have been apportioned among the several regions in a manner that is "fair". Any subset of the regions could reasonably ask the question: Do our shares represent a "fair" division among us considered as a *separate* group? Suppose that they believed that a different division of their pooled shares would be more fair. In that case it would be possible to substitute this different division for their initial apportionment to obtain a new apportionment among all that is "fairer" for the regions of the subset, and the same for all others. How could one then affirm that the initial division was fair to all? A rule that did this would surely be judged to be "incoherent"! The idea is captured in the slogan: "Any part of a fair division must be fair."

To be precise, consider an apportionment chosen by a method—a "global" apportionment—sum the seats it assigns to any subset of the regions, and consider the apportionment(s) obtained by applying the same method to redistribute this sum among the members of the subgroup (each of the latter set is a *local apportionment*). The method is *coherent* when two properties hold: (i) the shares assigned to each of the regions of the subset by the original (global) apportionment is a local apportionment and (ii) if there is another local apportionment among the regions of the subgroup, then another (global) apportionment of the method to all the regions is found as follows: substitute the shares in the local apportionment for what those regions have in the original apportionment.

For example, imagine an apportionment of the seats in the U.S. House of Representatives: if the method that is used yields an apportionment that gives 29 of the country's congressional seats to New York and 53 to California (as it did in accordance with the 2000 census), then surely the same method should divide the sum of 82 seats between New York and California in the same way. If the method *also* gives rise to a local apportionment that assigned 28 seats to New York and 54 to California, then replacing their shares in the initial apportionment with these should yield a second apportionment that belongs to the method. There would then be two possible apportionments of the House (a theoretical possibility unlikely to occur in practice). The concept of a coherent method[1] is quite general and is germane to many problems of fair division (Balinski, 2005), but it first arose in the context of the apportionment problem (Balinski and Young, 1982) where it is particularly important.

3.   From Divisions Between Two to Divisions Among All in Uni-Dimensional Apportionment

Given any apportionment problem (v, h), consider an apportionment obtained by a coherent rule. By definition, every pair of regions must share the seats they receive together h' in accordance with the rule applied to those regions when they are to be allocated h' seats. This immediately implies that knowing how to divide any number of seats between any *two* regions (meaning two regions having any number of inhabitants) suffices to completely determine a coherent rule. Deciding how to divide seats between only two regions is obviously an easier task than deciding how to divide seats among an arbitrary

---

[1] Coherence was earlier called "uniformity" and also "consistency."

number: this idea is pursued here to show how simple it is to extend vector apportionment to matrix apportionment.

Let Φ^2 be a method for dividing any number of seats between two regions (or parties), and Φ_d^2 be the divisor method based on d for dividing any number of seats between two regions (or parties). The methods Φ_d^2 enjoy two evident properties that will shortly be used. First, they are "monotone": if $(a_1, a_2) \in \Phi_d^2((v_1, v_2), h)$ and $(a_1', a_2') \in \Phi_d^2((v_1, v_2), h')$ for $h' > h$ then $a_1' \geq a_1$ and $a_2' \geq a_2$. Second, if $\{(a_1, a_2), (a_1', a_2')\} \subseteq \Phi_d^2((v_1, v_2), h)$ then $|a_1 - a_1'| \leq 1$ and $|a_2 - a_2'| \leq 1$: two apportionments for a same h can differ by at most 1.

Given any two apportionments a, b of a problem (v, h), consider $a - b$. It is a vector of integers summing to 0: so a may be obtained from b in a sequence of changes each of which transfers one seat from one party (or region) i where $b_i > a_i$ to another j where $a_j < b_j$: "local" change always involves just two parties (or regions).

A single property suffices to determine a divisor method.

PROPERTY 3 *A method Φ for vector problems is said to be **coherent with** Φ^2 if for every pair i, j*

$$a \in \Phi(v, h) \text{ implies } (a_i, a_j) \in \Phi^2((v_i, v_j), a_i + a_j). \text{ Moreover,}$$

$$(b_i, b_j) \in \Phi^2((v_i, v_j), a_i + a_j) \text{ implies } a' \in \Phi(v, h),$$

$$\text{where } a' = a \text{ except } a_i' = b_i, a_j' = b_j.$$

The really essential—and at first blush surprising—point about coherence is that Φ treats *every pair i, j* exactly as does Φ^2. In addition, if a change could be made that agrees with Φ^2 then it would yield another apportionment of Φ.

THEOREM 1 *The unique method Φ for vector problems that is coherent with Φ_d^2 is the divisor method Φ_d.*

Proof. The condition is necessary, since Φ_d is obviously coherent with Φ_d^2.

To see that the condition is sufficient, suppose Φ is any method that is coherent with Φ_d^2 and that $a \in \Phi(v, h)$. It is shown that $a \in \Phi_d(v, h)$. Choose $\lambda > 0$ such that $\sum_i [\lambda v_i]_d = h$, and let $b_i = [\lambda v_i]_d$, so $b \in \Phi_d(v, h)$, implying, in particular, that $(b_i, b_j) \in \Phi_d^2((v_i, v_j), b_i + b_j)$ for every pair i, j. The coherence of Φ with Φ_d^2 implies that $(a_i, a_j) \in \Phi_d^2((v_i, v_j), a_i + a_j)$ for every pair i, j. Suppose $b \neq a$. Then there exists a pair i, j for which $a_i > b_i$ and $a_j < b_j$. But this is impossible unless $a_i + a_j = b_i + b_j$ so $a_i = b_i + 1$, $a_j = b_j - 1$ and $\{(a_i, a_j), (b_i, b_j)\} \subseteq \Phi_d^2((v_1, v_2), a_i + a_j)$. The coherence of Φ_d with Φ_d^2 implies that substituting (a_i, a_j) for (b_i, b_j) in b yields an apportionment that belongs to $\Phi_d(v, h)$ which agrees with more components of a than did b. Repeating the same argument until a is obtained shows that $a \in \Phi_d(v, h)$, and so completes the proof.

So, if one wishes to verify that $a \in \Phi_d(v, h)$ it suffices to check that $(a_i, a_j) \in \Phi_d^2((v_i, v_j), a_i + a_j)$ for every pair i, j: "local" conditions determine the solution, just as in well-behaved optimization problems no possible local "improvement" implies the solution in hand is an optimum. The analogy with optimization carries further: the λ found in (1) may be viewed as a kind of "dual" variable which, once known, makes it possible to solve the problem by assigning the obvious value to each a_i *independently*, namely, by taking it to be a d-rounding of λv_i.

The parametric methods Φ_δ, $0 \le \delta \le 1$, each have a particularly simple closed-form formula for calculating how two regions divide any number of seats between them. The simplest—and most natural—of all is Webster's or Sainte-Laguّe's: the proportional share of each is rounded to the closest integer. For a positive real number x, suppose $x = n + r$, n integer and $0 \le r < 1$. Define $[x]_\delta = n$ if $r \le \delta$, and $[x]_\delta = n + 1$ if $r \ge \delta$ (so $[n + \delta]_\delta = n$ or $n + 1$). It is a simple exercise to show:

LEMMA 2 *Given a two-region problem* $((v_1, v_2), h)$, *let* $\bar{v}_k = v_k / (v_1 + v_2)$, $k = 1, 2$. *Then* $a_k = \left[\bar{v}_k (h + 2\delta - 1) \right]_\delta$, $k = 1, 2$ *(and if* $\bar{v}_k (h + 2\delta - 1)$ *has a remainder of exactly* δ *then one of the a's is rounded-up, the other is rounded-down).*

4. Matrix Apportionment: a Primer

A *matrix (or two-dimensional) apportionment problem* is a triple (v, r, c), where $v = (v_{ij}) \ge 0$ for $i = 1, \ldots, m$ and $j = 1, \ldots, n$ is a nonnegative matrix with no row or column of 0's, and $r = (r_1, \ldots, r_m) > 0$ and $c = (c_1, \ldots, c_n) > 0$ are vectors of integers whose sums are equal, $\sum r_i = \sum c_j = h$. An *apportionment* is a matrix $a = (a_{ij})$, where $a_{ij} \ge 0$ is integer valued, $\sum_j a_{ij} = r_i$ for all i and $\sum_i a_{ij} = c_j$ for all j. Matrix apportionment is a more recent problem where v_{ij} is the vote of party i's list in region j, r_i is the number of seats deserved by each party i (on the basis of the total vote of all of its lists $\sum_j v_{ij}$, for example) and c_j is the number of seats assigned to each region j (typically on the basis of its population): which one of the many possible apportionments a should be chosen?

In general, a *(matrix) method of apportionment* Φ selects a nonempty subset of apportionments $\Phi(v, r, c)$ for any problem (v, r, c).

When $v_{ij} = 0$ (or, more generally, when v_{ij} is less than some preset positive threshold) imposes that $a_{ij} = 0$, it may be that no apportionment exists. But this is rather unlikely. The example of figure 1 is typical of the only situations when none exists. The subset of regions (or columns) J that consists of the 4th through the 7th regions are to receive together a total of 8 seats (in general, $c(J) = \sum_J c_j$ seats). The subset of parties (or rows) I_J each of which received some votes (or more than the threshold of votes) from at least one of the regions

	1st	2nd	3rd	4th	5th	6th	7th	seats
Party 1	+	+	+	+	+	+	+	2
Party 2	+	+	+	+	+	+	+	5
Party 3	+	+	+	0	0	0	0	4
Party 4	+	+	+	0	0	0	0	6
Seats	4	2	3	1	1	2	4	

Fig. 1.   Example of votes that allows no feasible apportionment. (+ means any number of votes, 0 means no votes or too few to permit a seat.)

of J, namely, the parties 1 and 2, deserve a total of 7 seats (in general, $I_J = \{i : v_{ij} = +$ for some $j \in J\}$ having a total of $r(I_J) = \sum_{I_K} r_i$ seats). Thus the regions J are to have 8 seats but they can only fill them from candidates of the parties I_J who deserve 7 seats: clearly, there can be no apportionment in this case. If regions assigned 8 seats give all of their votes to parties that in total only deserve 7 seats, the total turnout in those regions must be abnormally low. There is a symmetric explanation. The set I consisting of parties 3 and 4 deserve $r(I) = 10$ seats; the set J_I consisting of those regions who gave some votes (or more than the threshold) to at least one of the lists of parties I, namely, regions 1, 2, and 3, are to receive $c(J_I) = 9$ seats. This is again clearly impossible since it asks that parties deserving 10 seats get them all from regions having only 9 seats, but also unlikely for in this case the turnout in the regions J_I must be abnormally high. Yet, as the following theorem shows, this is the only situation that can deny the existence of apportionments (its proof is easily deduced from duality in linear programming or from the min-cut, max-flow theorem of network flows).

THEOREM 3 *There exist apportionments if and only if $c(K) \leq r(I_K)$ for every subset of the regions (or columns) K.*

A problem that has apportionments will be said to be *feasible*.

A *(matrix) divisor method based on d* is, for any feasible problem, the set of apportionments:

$$\Phi_d(v, r, c) = \qquad\qquad\qquad (8)$$

$$\{a = (a_{ij}) : a_{ij} = [\lambda_i v_{ij} \mu_j]_d \text{ for } \lambda, \mu \text{ such that } \sum_j a_{ij} = r_i \text{ and } \sum_i a_{ij} = c_j\}.$$

Note, again, that $d(0) = 0$ implies $[\lambda_i v_{ij} \mu_j]_d \geq 1$ for every $\lambda_i v_{ij} \mu_j > 0$. This means that $\Phi_d(v, r, c)$ may be empty when $d(0) = 0$ despite the fact that the problem is feasible. In order for $\Phi_d(v, r, c)$ to be nonempty when $d(0) = 0$ there must exist an apportionment a that satisfies the row- and column- equations and also $a_{ij} \geq 1$ when $v_{ij} > 0$ and $a_{ij} = 0$ when $v_{ij} = 0$: call such problems *super-feasible*.

Properties of matrix methods of apportionment similar to those appealed to for vector methods of apportionment characterize the matrix divisor methods (Balinski and Demange, 1989a).

THEOREM 4 *(Balinski and Demange, 1989b). For any feasible matrix problem* (v, r, c) *there exist multipliers* λ, μ *and an* $a \in \Phi_d(v, r, c)$ *with* $a_{ij} = [\lambda_i v_{ij} \mu_j]_d$ *when* $d(0) > 0$; *and when* $d(0) = 0$ *the same is true if the problem is super-feasible. The multipliers are not unique, and there may be several apportionments in* Φ_d; *however, if there is more than one apportionment in* Φ_d, *all of them are obtained with a same set of multipliers.*

5.   From Divisions Between Two to Divisions Among All in Bi-Dimensional Apportionment

One property sufficed to determine a divisor method for uni-dimensional problems. For bi-dimensional problems two properties suffice.

Given an m by n matrix v, an m-vector $\lambda = (\lambda_1, \ldots, \lambda_m)$ and an n-vector $\mu = (\mu_1, \ldots, \mu_n)$, let $\lambda \circ v \circ \mu = (\lambda_i v_{ij} \mu_j)$; that is, the matrix obtained from v by multiplying its ith row by λ_i and its jth column by μ_j, for all i, j.

An essential property in vector apportionment is that a method (*any* method) should yield the same solutions to (v, h) and to $(\lambda v, h)$ for any scalar $\lambda > 0$: that is, how votes are scaled should make no difference. It came for free in the uni-dimensional case, but must be called upon in the bi-dimensional case. Since a party i (or row) deserves a fixed number of seats r_i, rescaling by multiplying its votes by $\lambda_i > 0$ should (as in the vector problem) change nothing; symmetrically, since a region j is assigned a fixed number of seats c_j, rescaling its votes by $\mu_j > 0$ should (as in the vector problem) change nothing as well.

PROPERTY 4 *A method* Φ *for matrix problems is said to be **proportional** if*

$$\Phi(v, r, c) = \Phi(\lambda \circ v \circ \mu, r, c) \text{ for every real } \lambda, \mu > 0.$$

Given any two apportionments a, b of a problem (v, r, c), consider $a - b$. It is a matrix of integers each of whose rows and columns sums to 0: so a may be obtained from b in a sequence of changes each of which transfers 1 seat from one to another entry of the matrix within a *simple cycle* C

$$
\begin{array}{ccccccccc}
i(1)j(1) & & i(2)j(2) & & \cdots & & i(k-1)j(k-1) & & i(k)j(k) & & i(1)j(1) \\
\downarrow & \nearrow & \downarrow & \nearrow & & \nearrow & \downarrow & & \downarrow & \nearrow & \\
i(1)j(2) & & i(2)j(3) & & \cdots & & i(k-1)j(k) & & i(k)j(1) & &
\end{array}
$$

$$(9)$$

for which $b_{i(s)j(s)} > a_{i(s)j(s)}$ and $b_{i(s)j(s+1)} < a_{i(s)j(s+1)}$ for $s = 1, \ldots, k$, where $k + 1$ is taken as 1, the indices $i(s)$ are different and so are the indices $j(s)$. The change decreases the $b_{i(s)j(s)}$ by 1 and increases the $b_{i(s)j(s+1)}$

by 1 in the cycle: "local" change always involves party-regions (i, j) that form a cycle. To simplify the description below, rename the "even" entries $(i(s), j(s)) = (s, s)$ and the "odd" entries $(i(s), j(s + 1)) = (s, s + 1)$ and $(i(k), j(1)) = (k, 1)$.

PROPERTY 5 *A method Φ for matrix problems is said to be* **coherent with** Φ^2 *if for any problem (v, r, c) there exists an equivalent problem (v', r, c), $v' = \lambda \circ v \circ \mu$, for which*

$$a \in \Phi \text{ implies } (a_{kl}, a_{st}) \in \Phi^2\big((v'_{kl}, v'_{st}), a_{kl} + a_{st}\big),$$

for every pair of indices $(k, l), (s, t)$. Moreover, suppose that for some simple cycle C as in (9) there is a b for which

$$(b_{ss}, b_{ss+1}) \in \Phi^2\big((v'_{ss}, v'_{ss+1}), a_{ss} + a_{ss+1}\big) \text{ for all } s(\text{mod } k), \text{ and}$$

$$(b_{s-1s}, b_{ss}) \in \Phi^2\big((v'_{s-1s}, v'_{ss}), a_{s-1s} + a_{ss}\big) \text{ for all } s(\text{mod } k).$$

Then

$$a' \in \Phi(v', r, c), \text{ where } a' = a \text{ except } a'_{ij} = b_{ij} \text{ for } (i, j) \in C.$$

Again, the really essential—and at first blush surprising—point about coherence is that Φ treats *every* pair $(k, l), (s, t)$ exactly as does Φ^2 relative to the equivalent problem (v', r, c). In addition, if a change could be made that agrees with Φ^2 with respect to the equivalent problem, then that would yield another apportionment of Φ. But a change in a matrix apportionment implies at least a change in a simple cycle C.

THEOREM 5 *The unique proportional method for matrix problems Φ that is coherent with Φ_d^2 is the divisor method Φ_d.*

Proof. The conditions are necessary since Φ_d is obviously proportional and coherent with Φ_d^2.

To see that the conditions are sufficient, suppose Φ is any proportional method that is coherent with Φ_d^2 and that $a \in \Phi(v, r, c)$. It will be shown that $a \in \Phi_d(v, r, c)$.

There exist $\lambda^b > 0$, $\mu^b > 0$ so that $b = (b_{ij}) \in \Phi_d(v, r, c)$, where $b_{ij} \in [\lambda_i^b v_{ij} \mu_j^b]_d$.

$a \in \Phi(v, r, c)$ and Φ coherent with Φ_d^2 implies there exist $\lambda^a > 0$, $\mu^a > 0$ so that $a_{ij} \in [\lambda_i^a v_{ij} \mu_j^a]_d$.

Suppose $a \neq b$. Then for some (i, j), $a_{ij} < b_{ij}$, and (i, j) belongs to a simple cycle C as in (8). Simplifying the notation again, let $\{(1, 1), (2, 2), \ldots, (s, s)\}$ be the even entries and $\{(1, 2), \ldots, (s-1, s), (s, 1)\}$ be the odd entries, so that $a_{ii} < b_{ii}$, $a_{ii+1} > b_{ii+1}$ for $i = 1, \ldots, s(\text{mod } s)$. Multiplying λ^a by

$\lambda_1^b / \lambda_1^a$ and dividing μ^a by the same amount changes nothing, so it may be assumed that $\lambda_1^a = \lambda_1^b$. Now notice that in general, if x, y are reals, $a \in [x]_d$ and $b \in [y]_d$, then $a > b$ implies $x \geq y$ and $a > b + 1$ implies $x > y$.

Begin the cycle C at $(1, 1)$ and follow it with the indices increasing. $\lambda_1^a = \lambda_1^b$ and $a_{11} < b_{11}$ implies $\mu_1^a \leq \mu_1^b$ (with strict inequality if $a_{11} + 1 < b_{11}$). Also, $\lambda_1^a = \lambda_1^b$ and $a_{12} > b_{12}$ implies $\mu_2^a \geq \mu_2^b$ (with strict inequality if $a_{12} > b_{12} + 1$). But $\mu_2^a \geq \mu_2^b$ and $a_{22} < b_{22}$ implies $\lambda_2^a \leq \lambda_2^b$ (with strict inequality if either $a_{22} + 1 < b_{22}$ or $\mu_2^a > \mu_2^b$). Continuing around the cycle, $\mu_s^a \geq \mu_s^b$ and $a_{ss} < b_{ss}$ implies $\lambda_s^a \leq \lambda_s^b$ (with strict inequality if either $a_{ss} + 1 < b_{ss}$ or $\mu_s^a > \mu_s^b$). But this means that $\lambda_s^a \leq \lambda_s^b$, $\mu_1^a \leq \mu_1^b$ and $a_{s1} > b_{s1}$, a contradiction *unless* the following holds: $\lambda_i^a = \lambda_i^b$, $\mu_i^a = \mu_i^b$ for $i = 1, \ldots, s$ *and* the differences between the values of the a's and b's in the cycle C are all exactly 1. But in this case there is a massive "tie". Defining $b' = b$, except that every b-entry in the cycle C is replaced by the corresponding value of a, another apportionment $b' \in \Phi_d(v, r, c)$ is obtained. Repeating the same argument until a is obtained shows that $a \in \Phi_d(v, r, c)$, and so completes the proof.

The analogy with optimization may be carried further here as well: the vectors λ, μ found in (7) may be thought of as "dual" variables which, once known, make it possible to solve the problem by assigning the obvious (or "greedy") value independently to each a_{ij}, namely, by taking it to be a d-rounding of $\lambda_i v_{ij} \mu_j$.

There can be no multipliers λ, μ that yield different solutions, showing that uni- and bi-proportional apportionments are, in essence, the same problem, and that a matrix apportionment treats every pair $(i, j), (k, l)$ fairly.

Acknowledgements

I am indebted to my friend and colleague, Friedrich Pukelsheim, for incisive constructive comments and corrections.

References

M. Balinski (2004), *Le suffrage universel inachevé*, Éditions Belin, Paris.

M. Balinski (2002), "Une 'dose' de proportionnelle: le système électorale mexicain," *Pour la science*, 58-59.

M. Balinski (2005), What is just? *American Mathematical Monthly*, **112**, 502-511.

M. Balinski and G. Demange (1989a), An axiomatic approach to proportionality between matrices, *Mathematics of Operations Research* **14**, 700-719.

M. Balinski and G. Demange (1989b), Algorithms for proportional matrices in reals and integers, *Mathematical Programming* **45**, 193-210.

M. Balinski and S. Rachev (1997), Rounding proportions : methods of rounding, *The Mathematical Scientist* **22**, 1-26.

M. Balinski and V. Ramírez (1997), Mexican electoral law: 1996 version, *Electoral Studies* **16**, 329-340.

M. Balinski and V. Ramírez (1999a), Mexico's 1997 apportionment defies its electoral law, *Electoral Studies* **18**, 117-124.

M. Balinski and V. Ramírez (1999b), Parametric methods of apportionment, rounding and production, *Mathematical Social Sciences* **37**, 107-122.

M.L. Balinski and H.P. Young (1982), *Fair Representation: Meeting the Ideal of One Man, One Vote*, Yale University Press, New Haven; 2nd ed., Brookings Institution Press, Washington, D.C., 2001.

F. Pukelsheim and C. Schuhmacher (2004), Das neue Zürcher Zuteilungsverfahren für Parlamentswahlen, *Aktuelle Juristichische Praxis* **5**, 505-522.

P. Zachariassen and Martin Zachariassen (2005), A comparison of electoral formulae for the Faroese Palriament (The Løgting), Technical report, Náttúruvísindadeildin, Føroya, 1 August 2005.

Minimum Total Deviation Apportionments

Paul H. Edelman
Department of Mathematics and the Law School, Vanderbilt University, Nashville

Abstract   This note presents an algorithm for computing the minimum total deviation apportionment. Some properties of this apportionment are also explored. This particular apportionment arises from the jurisprudential concern that total deviation is the appropriate measure for the harm caused by malapportionment of the United States House of Representatives.

Keywords: Apportionment, House of Representatives, districting, one person - one vote.

1.   Introduction

The goal of this paper is to revive interest in evaluating methods of apportionment based on the objective functions that they optimize rather than their intrinsic axiomatic properties. The latter approach is certainly the dominant one as evidenced by such texts as (Balinski and Young, 2001) and (Saari, 1994). Nevertheless there are circumstances for which this may not be the best approach. I have argued elsewhere (Edelman, to appear) that the case of the apportionment of the United States House of Representatives is exactly such a circumstance. Subsequent to the "one person, one vote" rulings of the mid-1960's, the United States Supreme Court has adopted the measure of *total deviation* to quantify the harm resulting from unequal voting district sizes. Once having established a measurement of the harm, the Court should require that any apportionment do what it can to mitigate that harm. This implies that any method of apportionment should look to minimize the total deviation.

As it happens there are two papers, both pre-dating "one person, one vote," investigating methods of apportionment that minimize total deviation. The first, by Burt and Harris (1963), argued in favor of apportioning the House of Representatives so as to minimize total deviation on equitable principles and presented an algorithm using dynamic programming to find such an apportionment. This paper has been cited a number of times in the literature.

A year later and in the same journal, Gilbert and Schatz (1964) published a response to Burt and Harris. Their rebuttal made three arguments: First, the equitable arguments in favor of minimizing total deviation were not convincing;

second, there may be many apportionments that minimize total deviation, and last, that the algorithm provided by Burt and Harris to produce the minimizing apportionment was unduly complicated. They provided quite an elegant algorithm, which I will present subsequently.

Oddly, Gilbert and Schatz's article seems to have escaped notice. As far as I know it has never been cited. Yet it contains some quite lovely ideas. A secondary purpose of this paper is to present the ideas of Gilbert and Schatz in a contemporary setting so they will get the attention that I think they deserve.

This paper is organized as follows: The next section presents the necessary background from the theory of apportionment. It is necessarily brief, and I will rely on the reader to have a basic familiarity with the techniques. Section 3 presents an algorithm to compute the minimum total deviation (mtd) apportionments. The method presented is due to Gilbert and Schatz (1964) although I have streamlined the presentation and proofs. The next three sections discuss technical issues associated with mtd apportionments. Section 4 confronts the problem of multiple mtd apportionments, Section 5 discusses bias and Section 6 examines the Alabama paradox. Section 7 is a brief conclusion.

2.   Preliminaries

In this section I will introduce the necessary terminology. Since I am primarily interested in the apportionment of the United States House of Representatives, I will phrase the apportionment problem in terms of assigning seats to states. Assume that there are s states and let $\mathbf{p} = (p_1, p_2, \ldots, p_s)$ be the state populations. For h a positive integer we call $\mathbf{a} = (a_1, a_2, \ldots, a_s)$ an h-apportionment if $\sum a_i = h$. We will refer to a_i as the number of seats that state i receives. I will assume throughout that the state populations are generic in the sense that

$$\frac{p_i}{j} \neq \frac{p_k}{l}$$

for $1 \leq i, k \leq s$ and for all positive integers $1 \leq j, l \leq s$.

Given $\mathbf{p}$ and h-apportionment $\mathbf{a}$ let

1 $Max(\mathbf{p}, \mathbf{a}) = \max_i \frac{p_i}{a_i}$

2 $Min(\mathbf{p}, \mathbf{a}) = \min_i \frac{p_i}{a_i}$, and

3 $TD(\mathbf{p}, \mathbf{a}) = \max_{i,j}\{\frac{p_i}{a_i} - \frac{p_j}{a_j}\} = Max(\mathbf{p}, \mathbf{a}) - Min(\mathbf{p}, \mathbf{a})$.

Thus, $Max(\mathbf{p}, \mathbf{a})$ is largest population/seat ratio among the states, $Min(\mathbf{p}, \mathbf{a})$ is the smallest such value, and $TD(\mathbf{p}, \mathbf{a})$, the *total deviation* of the apportionment, is the gap between these two values.

Two methods of apportionment will be of particular importance in this paper. The first is the Adams method, which can be described in the following way (Balinski and Young, 2001, page 142): Given s states with populations $\mathbf{p}$, and a house of size h, $h \geq s$, we let $Adams(\mathbf{p}, h)$ be the h-apportionment given recursively by

1   $Adams(\mathbf{p}, s) = (1, 1, \ldots, 1)$,

2   Let $\mathbf{A} = Adams(\mathbf{p}, h - 1)$. If t is the state so that $\frac{p_t}{a_t} = Max(\mathbf{p}, \mathbf{A})$
    then define

$$Adams(\mathbf{p}, h)_i = \begin{cases} A_i + 1, & \text{if } i = t; \\ A_i, & \text{otherwise.} \end{cases}$$

Note that from the definition we have that $Max(\mathbf{p}, Adams(\mathbf{p}, h))$ is strictly decreasing as a function of h.

LEMMA 1 *The Adams apportionment* $Adams(\mathbf{p}, h)$ *minimizes* $Max(\mathbf{p}, \mathbf{a})$ *over all h-apportionments* $\mathbf{a}$.

Proof. This fact is noted in (Balinski and Young, 2001, page 104) without proof. For the sake of completeness I include one here. Let $Adams(\mathbf{p}, h) = \mathbf{A}$ and suppose there is an h-apportionment $\mathbf{a}$ so that

$$\frac{p_i}{a_i} = Max(\mathbf{p}, \mathbf{a}) < Max(\mathbf{p}, \mathbf{A}) = \frac{p_j}{A_j}.$$

It follows that $a_j > A_j$ and, since both $\mathbf{a}$ and $\mathbf{A}$ are h-apportionments, there must be some k so that $a_k < A_k$. Since the Adams h-apportionment assigns more seats to state k than $\mathbf{a}$ does, it follows that for some $h' < h$ we have $\frac{p_k}{a_k} = Max(\mathbf{p}, Adams(\mathbf{p}, h'))$. Since $Max(\mathbf{p}, Adams(\mathbf{p}, h))$ is strictly decreasing as a function of h, $\frac{p_k}{a_k} > \frac{p_j}{A_j}$ which contradicts the assumption that $\frac{p_i}{a_i} = Max(\mathbf{p}, \mathbf{a})$. ∎

If $h' > h$ and $\mathbf{a}$ and $\mathbf{a}'$ are h- and h'-apportionments, respectively, I will say that $\mathbf{a}'$ is an h'-*extension* of $\mathbf{a}$ if $\mathbf{a}' \geq \mathbf{a}$, i.e., $a'_k \geq a_k$ for all $1 \leq k \leq s$. The following lemma helps to illustrate this idea and will prove useful in the next section.

LEMMA 2 *Let* $\mathbf{A} = Adams(\mathbf{p}, h)$. *If* $\mathbf{a}$ *is an apportionment with*

$$Max(\mathbf{p}, \mathbf{a}) = Max(\mathbf{p}, \mathbf{A})$$

then $\mathbf{a}$ *is an h'-extension of* $\mathbf{A}$ *for some* $h' \geq h$.

Proof. Let $\frac{p_k}{A_k} = Max(\mathbf{p}, \mathbf{A})$. Because the Adams method always gives priority to the state with the largest ratio of population to seats, we know that $\frac{p_l}{A_l-1} > \frac{p_k}{A_k}$ for all $l \neq k$, since the Adams method gave state l its A_l^{th} seat before k gets its A_k^{th}. Thus, $Max(\mathbf{p}, \mathbf{a}) = Max(\mathbf{p}, \mathbf{A})$ implies that $a_l \geq A_l$ for all l and thus $\mathbf{a}$ is an extension of $\mathbf{A}$. ∎

If $\mathbf{a}$ is an h-apportionment, the h'-Jefferson extension of $\mathbf{a}$, $JExt(\mathbf{a}, h')$ is defined recursively by:

1 $JExt(\mathbf{a}, h) = \mathbf{a}$,

2 Let $\mathbf{J} = JExt(\mathbf{a}, h' - 1)$. If t is the state so that

$$\frac{p_t}{J_t + 1} = Max_i\{\frac{p_i}{J_i + 1}\}$$

then define

$$JExt(\mathbf{p}, h')_i = \begin{cases} J_i + 1, & \text{if } i = t; \\ J_i, & \text{otherwise.} \end{cases}$$

It is clear that $Jeff(\mathbf{p}, h) = JExt(\mathbf{0}, h)$, where $\mathbf{0} = (0, \ldots, 0)$, is just the usual Jefferson h-apportionment (Balinski and Young, 2001, page 142).

LEMMA 3 *If $\mathbf{a}$ is an h-apportionment, and $h' \geq h$, then $JExt(\mathbf{a}, h')$ maximizes $Min(\mathbf{p}, \mathbf{a}')$ over all h'-extensions of $\mathbf{a}$.*

Proof. An essentially equivalent fact is stated in (Balinski and Young, 2001, page 104) without proof. For completeness I include one here.

Let $\mathbf{J} = JExt(\mathbf{a}, h')$. Suppose that $\mathbf{a}'$ is another h'-extension of $\mathbf{a}$ for which

$$\frac{p_i}{a'_i} = Min(\mathbf{p}, \mathbf{a}') > Min(\mathbf{p}, \mathbf{J}) = \frac{p_k}{J_k}.$$

It follows that $a'_k < J_k$, and since both $\mathbf{J}$ and $\mathbf{a}'$ are h'-extensions, there must be some l so that $a'_l > J_l$. We also know that $a_k < J_k$ and so state k received at least one more seat in $JExt$ than in $\mathbf{a}$. From the definition of $JExt$, then, we know that

$$\frac{p_k}{J_k} > \frac{p_l}{J_l + 1} \geq \frac{p_l}{a'_l} > \frac{p_i}{a'_i}$$

which is a contradiction. ∎

There is but one last piece of notation required. Suppose that $\mathbf{a}$ is an h-apportionment and let k be a state. By $\mathbf{a}|_k$ I mean the $(h - a_k)$-apportionment for the states with k removed.

3. Minimum Total Deviation Apportionment

In this section I will present the algorithm for finding an apportionment that minimizes the total deviation function $TD(\mathbf{p}, \mathbf{a})$. This algorithm first appeared in (Gilbert and Schatz, 1964) and I have done little to improve it other than to update the terminology and streamline the proof. Their idea is quite clever and deserves to be more widely known. The key to the construction is to begin an apportionment using the Adams method, but extend it using the Jefferson extension. Since the Adams method minimizes $Max(\mathbf{p}, \mathbf{a})$ and the Jefferson extension maximizes $Min(\mathbf{p}, \mathbf{a})$ combining the two methods results in minimizing the gap $TD(\mathbf{p}, \mathbf{a})$.

Let $\mathbf{p}$ be the set of state populations as before, and suppose we want to find the h'-apportionment that minimizes the total deviation. For $h, h' \geq h \geq s$, let $\mathbf{A}^h = Adams(\mathbf{p}, h)$. If k is the state so that

$$\frac{p_k}{A_k^h} = Max(\mathbf{p}, \mathbf{A}^h)$$

let $\mathbf{J}^h$ be the h'-apportionment obtained by taking $JExt(\mathbf{A}^h|_k, h' - A_k^h)$ and then assigning A_k^h seats to state k. That is, $\mathbf{J}^h$ is obtained by assigning the first h seats using Adams method, setting aside the state which maximizes the population/seat ratio and then extending the rest of the apportionment using Jefferson's method. Thus, $\mathbf{J}^h$ is an h'-apportionment of $\mathbf{p}$ for every $h, h' \geq h \geq s$.

THEOREM 4 *The minimum of $TD(\mathbf{p}, \mathbf{a})$ over all h'-apportionments is equal to*

$$\min_{\{h \mid h' \geq h \geq s\}} TD(\mathbf{p}, \mathbf{J}^h).$$

That is, the minimum of $TD(\mathbf{p}, \mathbf{a})$ over all h'-apportionments is achieved by one of the h'-apportionments in the set $\{\mathbf{J}^h\}$.

Proof. Suppose that $\mathbf{a}'$ is an h'-apportionment that achieves the minimum of $TD(\mathbf{p}, \mathbf{a})$. Let

$$\frac{p_k}{a_k'} = Max(\mathbf{p}, \mathbf{a}').$$

Since the Adams h'-apportionment minimizes $Max(\mathbf{p}, \mathbf{a})$ over all h'-apportionments, it must be true that for some $h \leq h'$ we have

$$\frac{p_k}{a_k'} = Max(\mathbf{p}, Adams(\mathbf{p}, h)).$$

By Lemma 2 we know that $\mathbf{a}'$ is an h'-extension of $Adams(\mathbf{p}, h)$. Thus $\mathbf{a}'|_k$ is an extension of $Adams(\mathbf{p}, h)|_k$ and so, by Lemma 3, $Min(\mathbf{p}, \mathbf{a}') \leq Min(\mathbf{p}, \mathbf{J}^h)$. Thus,

$$TD(\mathbf{p}, \mathbf{a}') \geq TD(\mathbf{p}, \mathbf{J}^h)$$

and the theorem is proved. ∎

This idea of starting an apportionment using one standard method and then extending it using a different one is interesting and understudied. It is also more subtle than it might seem at first glance. One might think that the above construction done in the opposite order, i.e., start an apportionment using the Jefferson method and then extend it using Adams, would result in the same outcome, but it need not. For while no extension of an apportionment can increase $Max(\mathbf{p}, \mathbf{a})$, most extensions will result in decreasing $Min(\mathbf{p}, \mathbf{a})$. Thus following Jefferson with Adams will almost surely result in losing control of $Min(\mathbf{p}, \mathbf{a})$ and no claim similar to Theorem 4 will be true.

4.   A Multiplicity of Minima

An unfortunate aspect of total deviation is that minimizing apportionments need not be unique. This was observed first by Gilbert and Schatz (1964) who provided an example based on a modification of the 1960 US census data. The existence of such examples was, for them, a reason to disqualify minimizing total deviation as a means of choosing an apportionment. I have argued otherwise, elsewhere (Edelman, to appear). Nevertheless, this is an unusual aspect of total deviation which is worth considering further.

If the House of Representatives were apportioned using a total deviation minimizing method, then two of the twenty-two apportionments would not have been unique. In both 1810 and 1840 there were multiple apportionments that had the same minimum total deviation. Table 1 lists the 14 different apportionments for the House in 1810 which achieve the minimum.

Table 1.   1810 Minimum Total Deviation Apportionments

State	Population	1	2	3	4	5	6	7	8	9	10	11	12	13	14
New York	953043	26	26	26	26	26	26	27	27	27	27	25	25	25	25
Virginia	817615	22	22	22	23	23	23	22	22	22	23	22	23	23	23
Pennsylvania	809773	22	23	23	22	22	23	22	22	23	22	23	22	23	23
Massachusetts	700745	20	19	20	19	20	19	19	20	19	19	20	20	19	20
North Carolina	487971	14	14	13	14	13	13	14	13	13	13	14	14	14	13
Kentucky	374287	10	10	10	10	10	10	10	10	10	10	10	10	10	10
South Carolina	336569	9	9	9	9	9	9	9	9	9	9	9	9	9	9
Maryland	335946	9	9	9	9	9	9	9	9	9	9	9	9	9	9
Connecticut	261818	7	7	7	7	7	7	7	7	7	7	7	7	7	7
Tennessee	243913	7	7	7	7	7	7	7	7	7	7	7	7	7	7
New Jersey	241222	7	7	7	7	7	7	7	7	7	7	7	7	7	7
Ohio	230760	6	6	6	6	6	6	6	6	6	6	6	6	6	6
Vermont	217895	6	6	6	6	6	6	6	6	6	6	6	6	6	6
New Hampshire	214460	6	6	6	6	6	6	6	6	6	6	6	6	6	6
Georgia	210346	6	6	6	6	6	6	6	6	6	6	6	6	6	6
Rhode Island	76888	2	2	2	2	2	2	2	2	2	2	2	2	2	2
Delaware	71004	2	2	2	2	2	2	2	2	2	2	2	2	2	2

It is interesting to note that the Hamilton, Webster, Dean, and Hill methods all give the same apportionment in this case and that apportionment, number 4 in Table 1, is a mtd apportionment. The population data for the 1840 census, which also has multiple *mtd* apportionments has a similar property; Hamilton,

Webster, Dean and Hill all agree, although this time that apportionment is not *mtd*. This suggests that one might be able to say more about population data that produces multiple *mtd* apportionments. What can one say about situations for which the *mtd* apportionment is unique? To begin, observe that the genericity assumption that the population/seat ratios are all distinct does not exclude that the *differences* of the ratios are distinct. This opens the door for two apportionments to have the same total deviation by chance. For example, consider three states, $\mathbf{A}, \mathbf{B}$ and $\mathbf{C}$, with populations $4704, 2076$ and 539, respectively. One can check that the minimum total deviation is achieved by two different apportionments, $(8, 3, 1)$ and $(7, 4, 1)$. In the first apportionment state $\mathbf{B}$ has the largest population/seat ratio of 692, $\mathbf{C}$ has the smallest at 539 for a total deviation of 153. In the second apportionment, state $\mathbf{A}$ is largest (672) and $\mathbf{B}$ is smallest (519) for the same total deviation.

I will call a set of populations *hyper-generic* if, not only are the population/seat ratios distinct (for all house sizes suitably small), but the differences of population/seat ratios are also distinct. For hyper-generic populations, two apportionments can have the same total deviation only if the states achieving the maximum and minimum population/seat ratios are the same in each apportionment and the differences occur in the distribution of seats among the other states. The data from the 1810 census illustrates this situation, where Ohio has the largest population/seat ratio and New Jersey the smallest. The variation in the apportionments comes from reallocating the seats among some of the other states in such a way that their population/seat ratios stay within the range established by Ohio and New Jersey. The question then becomes when such an internal reallocation is possible. There are a few situations in which we can say something concrete about whether a *mtd* apportionment is unique:

LEMMA 5 *If the state populations are hyper-generic, and the Adams apportionment minimizes total deviation, then it is the unique apportionment that minimizes total deviation.*

Proof. In order for there to be another *mtd* apportionment, one must be able to reallocate a seat from one state to another. But in the Adams apportionment reducing the number of seats to any state will increase its population/seat ratio above that of the current maximum and thus increase the total deviation. ∎

In the following lemmas I will use the notation from Section 3. Recall that $\mathbf{J}^h$ is the h'-apportionment obtained from $Adams(\mathbf{p}, h)$ by taking $JExt(\mathbf{A}^h|_k, h'-A_k^h)$, where k is the state with the largest population/seat ratio in $Adams(\mathbf{p}, h)$, and then assigning A_k^h seats to state k.

LEMMA 6 *Suppose the state populations* $\mathbf{p}$ *are hyper-generic and* $\mathbf{J}^h$ *is a* mtd *h'-apportionment. This is the unique* mtd *apportionment if*

$$JExt(\mathbf{A}^h|_k, h' - A_k^h) = Jeff(\mathbf{p}|_k, h' - A_k^h).$$

That is, if the Jefferson extension agrees with the actual Jefferson apportionment, then the mtd *apportionment will be unique.*

Proof. It follows from Lemma 3 that any $h' - A_k^h$-apportionment **a** for the states $\mathbf{p}_k$ will have

$$Min(\mathbf{p}_k, \mathbf{a}) < Min(\mathbf{p}_k, Jeff(\mathbf{p}|_k, h' - A_k^h)) = JExt(\mathbf{A}^h|_k, h' - A_k^h).$$

Thus $\mathbf{J}^h$ must be the unique *mtd* apportionment. ∎

LEMMA 7 *Suppose that* $\mathbf{J}^h$ *is a* mtd h'-*apportionment, that it differs from* $Adams(\mathbf{p}, h)$ *in at least 2 states, and* $JExt(\mathbf{A}^h|_k, h' - A_k^h) \neq Jeff(\mathbf{p}|_k, h' - A_k^h)$. *Then the* mtd h'-*apportionment is not unique.*

Proof. That

$$JExt(\mathbf{A}^h|_k, h' - A_k^h) \neq Jeff(\mathbf{p}|_k, h' - A_k^h)$$

implies that there is some state $j \neq k$, with seat allocation J_j^h, so that

$$\frac{p_j}{J_j^h + 1} > Min(\mathbf{p}, \mathbf{J}^h).$$

Moreover, since there are at least two states on which $\mathbf{J}^h$ differs from $Adams(\mathbf{p}, h)$, there must be a state i different from both j and k so that

$$\frac{p_i}{J_i^h - 1} < Max(\mathbf{p}, \mathbf{J}^h)$$

and hence the transfer of a seat from state j to state i will result in an h'-apportionment with the same total deviation as $\mathbf{J}^h$. ∎

These three lemmas leave just a little uncertainty about the nature of the *mtd* apportionments that are not unique. The remaining case is if $\mathbf{J}^h$, the *mtd* apportionment, differs from $Adams(\mathbf{p}, h)$ in only one state. Such apportionments may or may not be unique depending on the existence of a second state to which a seat can be added without decreasing $Min(\mathbf{p}, \mathbf{J}^h)$. Either possibility can arise.

5.   Bias

A traditional concern in apportionment is whether there is an inherent bias in the method with respect to the size of the state. It is well-established (Balinski and Young, 2001, Chapter 9) that among standard methods, Hamilton and Webster are unbiased while Adams is biased toward small states and Jefferson is biased toward large ones. What can one say about the bias inherent in the *mtd* apportionment?

As noted in the previous section, in 20 of the 22 apportionments of the United States, the *mtd* apportionment was the same as the Adams apportionment. Thus, one might conclude that the *mtd* apportionment has a bias to small states. The difficulty with this line of reasoning is that, as shown previously, there may be a multiplicity of *mtd* apportionments and in those situations, there may be little or no bias. For example, as previously noted, among the *mtd* apportionments for the 1810 census data is the apportionment that agrees with the Hamilton, Webster, Hill and Dean methods.

So, to prove anything conclusively we would need more detailed information on two aspects of *mtd* apportionments; first, how often are there multiple *mtd* apportionments, and second, can we choose among multiple *mtd* apportionments in such a way as to minimize the resulting bias overall. The results in the previous section are but a small step in the first direction. The second is totally unresearched.

6.   Alabama Paradox

A method of apportionment is said to exhibit the Alabama paradox if an increase in the size of the house may result in the decrease in the number of seats allocated to a state. It is well-known that the divisor methods do not exhibit the paradox, while the Hamilton method does. In what way does the *mtd* apportionment behave?

Since *mtd* apportionments may not be unique one must be careful in how this problem is phrased. There is no question that if the *mtd* apportionments are chosen injudiciously the Alabama paradox may result. Consider the apportionment problem (taken from Figure 1) that consists of states New York, Virginia, Pennsylvania, New Jersey, and Ohio, with populations 953043, 817615, 809773, 241222, and 230760, respectively. One *mtd* 83-apportionment is 26, 22, 22, 7, and 6 seats for each state, respectively. A *mtd* 84-apportionment is 25, 23, 23, 7, and 6. So this pair of apportionments exhibits the Alabama paradox. On the other hand, 25, 23, 22, 7, and 6 is also a *mtd* 83-apportionment, which would show no Alabama paradox. It is also true that 26, 23, 22, 7, and 6 is a *mtd* 84-apportionment. So, by making an appropriate choice among the *mtd* apportionments one need not have the Alabama paradox manifest itself in this instance.

Can one always avoid the Alabama paradox in this way? I don't know. Balinski and Young (Balinski and Young, 2001, Proposition 3.9) assert that apportionments can exhibit the Alabama paradox, but they give no specific example. This leaves it unclear whether they were referring to the phenomenon just discussed or a more robust example in which the Alabama paradox is unavoidable.

7.   Conclusion

What I have presented here is a method of apportionment designed to minimize total deviation, a particular measure of harm in malapportionments. It is the measure of harm that has been recognized by the United States Supreme Court. While this method has less desirable behavior than standard methods from an axiomatic point-of-view, that does not mean that it is inappropriate for certain purposes. And it certainly does not mean that it is not an interesting method worth studying further.

References

Oscar R. Burt and Curtis C. Harris, Jr., *Apportionment of the U. S. House of Representatives: A minimum range, integer solution, allocation problem*, Oper. Res. **11**(1963), 648–652.

Michel L. Balinski and H. Peyton Young, FAIR REPRESENTATION, MEETING THE IDEAL OF ONE MAN, ONE VOTE, 2nd ed. Brookings Institution Press, Washington, DC, 2001.

Paul H. Edelman, *Getting the math right: Why California has too many seats in the House of Representatives*, Vanderbilt Law Review, to appear.

E. J. Gilbert and J. A. Schatz, *An ill-conceived proposal for apportionment of the U. S. House of Representatives*, Oper. Res. **12**(1964), 768–773.

Donald G. Saari, GEOMETRY OF VOTING, Springer-Verlag, New York, 1994.

Comparison of Electoral Systems:
Simulative and Game Theoretic Approaches

Vito Fragnelli[1], Guido Ortona[2]

[1]Dipartimento di Scienze e Tecnologie Avanzate, Università del Piemonte Orientale

[2]Dipartimento di Politiche Pubbliche e Scelte Collettive, Università del Piemonte Orientale

Abstract   Simulation may be a useful tool to address some basic problems concerning the choice of the electoral system. A case study is analyzed as an example. The utility of including power indices is discussed. A simulation program is illustrated.

Keywords: Simulation, Electoral Systems, Weighted Majority Games.

1.   Introduction

Simulation may be very useful in the comparative assessment of electoral systems; actually, it is difficult to imagine a field where the simulative approach may be more effective. There are two reasons for that. The first is that the real world feature that must be simulated is very simple - a set of preferences. The assessment of the relative performance of electoral systems requires a set of preferences, but is entirely *downstream* of the reasons that produced a set or another one. A 'virtual' case of a society that uses perfect proportionality and where there are some major parties and a cohort of minor ones provides nearly the same information offered by an analogous real-world case (pre-reform Italy, in this paper).

The second reason is even more compelling, and possibly less obvious, at least for non-social scientists. While the virtual *set* of preferences is nearly as informative as a real one, the *single* virtual subject is *identical* to a real one. According to the basic theorems of choice (Arrow's and May's), and more generally to the basic individualistic paradigm of social sciences, no preference must be privileged. Hence, there is no reason to ask *why* a given subject provided a given choice. The entire process of evaluating the result of his/her and others' combined choices is again downstream. In other terms, in this field the virtual subjects include *all* the relevant features of the real ones.

In this paper we argue in favor of the simulative approach for the evaluation of electoral systems. In Section 2 we present a simple empirical rule that allows choosing among electoral systems, in the line discussed in (Fragnelli et al., 2005). As we will see, the procedure requires the decision-maker to explicit its preferences; consequently, the choice rule is not affected by the theorem of Arrow (and cannot aspire to pinpoint the system *objectively* preferable). Section 3 contains an example obtained employing ALEX4, the improved version of a new and more powerful simulation program, announced (as ALEX3), but not used in (Fragnelli et al., 2005)[1]. Section 4 extends the discussion to power matters, proposing tools and methodologies for defining better indices. Conclusions are in Section 5. Technicalities are in appendices.

2.   The Choice of the Optimal Electoral System

The choice of the best electoral system affects a lot of facets of the political process. (Ortona, 2002) provides a list with some twenty items, arguably not complete. Fortunately, however, there is a general agreement that the efficiency in representing electors' will (*representativeness, R*) and the effect on the efficiency of the resulting government (*governability, G*) are of paramount relevance [2],[3]. There are at least two good reasons to privilege R and G.

First, to summon the representatives and to form a government are the basic aims of a Parliament (bar, obviously, to make laws). Possible pitfalls of other dimensions may be managed in other moments of the political process, but this is not the case for representativeness and governability, if we admit the sovereignty of the voters in choosing their representatives and that of the representatives in choosing the government. In addition, it is sensible to think that other dimensions are lexicographic with respect to them [4]. If this is so, the results obtained with reference to R and G will keep their validity irrespective of the dimensions judged relevant.

R and G may be evaluated through the assessment of plausible (albeit arbitrary) numerical indicators. The ones used in our simulations are briefly described in Appendix B (for further details, see (Bissey et al., 2004)). We will label them r and g, respectively [5]. The range of both is the interval 0-1.

---

[1] The most relevant additional features are the consideration of new electoral systems (including the Single Transferable Vote) and the possibility to define individually the virtual voters.

[2] See (Bowler et al., 2005).

[3] A more detailed characterization of both R and G and of the related trade-off (R is likely to increase with the number of parties and G to decrease) is provided in Appendix B, through the definition of the indices employed to assess them. For a broader discussion, see (Ortona, 1998) and (Bissey et al., 2004).

[4] Note however that the method outlined here may be extended to further dimensions, provided that suitable indices are available.

[5] A slightly different version of these indices has been employed also in (Ortona, 1998).

Results for different electoral systems, *referring to a single case*, may consequently be computed out. There are three possibilities. First, the values of both r and g of a given system may be greater than those of *all* other systems considered; we define that system *dominant*, and it is obviously the best. Unfortunately, this system is very likely not to exist, given the trade-off between the two dimensions. Second, the values of both r and g for a given system may be lower than those of *another* one. We define that system *dominated*, and it may safely be excluded: no need to consider system X, if system Y, better on both dimensions, is available. Third, systems may be neither dominant nor dominated, i.e. all of them are Pareto optimal, like (usually) plurality voting and proportional representation in real world. We label these systems *alternative*. Obviously, the rule we look for is useful only if it allows choosing among alternative systems. Note that there may be at most one (strongly) dominant system, while the dominated systems can be more than one.

In principle, to compare different electoral systems, we need voting results for different systems: a majoritarian vote, a proportional vote, a list of voters' ordered preferences for Condorcet voting, and so on. It is usually impossible to collect these data from real world. But given a set of virtual electors, each with her/his preferences, it is perfectly possible to produce them. Given the votes, every system considered will produce a potential Parliament, and each Parliament will have a pair of values of r and g. If a system will result as dominant, it is the good one; but, as we noticed, this result is very unlikely, given the trade-off between the two dimensions. Apparently, what we need to compare them is a social utility function (*SUF*) - admittedly a quite formidable requirement, to say little. Actually, we may be satisfied with something less.

Let us admit the *SUF* for representativeness and governability to be a typical Cobb-Douglas function in g and r, $U = Kg^a r^b$, where K is a suitable constant. We choose this form not only for its simplicity and versatility, but also for the meaning of a and b, the partial elasticity of U with reference to g and r, respectively; as we will see, this provides a meaningful characterization of the choice rule. Now consider two non-dominated systems, X and Y. We may write that:

$$X \succ Y \iff Kg_X^a r_X^b > Kg_Y^a r_Y^b \tag{1}$$

where $X \succ Y$ means that system X is preferred to system Y.

Let $p = \dfrac{a}{b}$. It is easy to obtain that condition (1) reduces to:

$$p > \frac{\ln \frac{r_Y}{r_X}}{\ln \frac{g_X}{g_Y}} \tag{2}$$

supposing that X refers to the system with the higher value of g.

REMARK 1 *The comparison of systems is strongly influenced by the actual scaling of the indices with respect to each other. This inconvenience is reduced* *by the choice of a multiplicative utility function. Suppose that a decision-maker thinks that g should be given more (less) relevance, and that the increase (decrease) of relevance may be established through the attachment to g of a multiplicative constant > 1 (< 1). This procedure leaves the choice rule unaffected.*

It is important to notice that the ratio of the elasticities, p, can be seen as the *price* in terms of a relative decrease of r that the community accepts to pay for a given relative increase of g. If, for instance, we have $p = 2$, it is worthwhile to accept a 2% reduction of r to gain a 1% increase of g (but for the approximation due to the use of differentials). In general, if an increase in g is valued more than the same increase in r, then $p > 1$, and vice versa [6].

The only a priori information we need to assess the fulfillment of the condition, is the value of p. We argue that this parameter may actually be provided by the political system. Several procedures may be adopted, as discussed in Ortona (2005).

Equation (2) allows for binary comparisons of non-dominated electoral systems, and hence for finding out the Condorcet winner. The winner is the best system [7].

Alternatively, we may trace indifference curves and pick the system that lies on the higher curve. This procedure allows for a graphical individuation of the preferred system. For details, see (Fragnelli et al., 2005).

3.   An Example

In this section we provide an example that mirrors the actual Italian case.

The input is a representative survey of electoral preferences of Italian citizens collected by the Osservatorio del Nord Ovest of the Università di Torino in the first quarter of 2004. The simulation program described in Appendix A provides the data of Table 1 and Figure 1.

The choice set may be considerably reduced through the exclusion of systems that are *dominated* or *weakly dominated*. This criterion leaves us with the ten systems labeled 2, 3, 4, 6, 8, 11, 17, 18, 20, 21.

An elicitation procedure implemented with 80 students at the Laboratorio di Economia Sperimentale e Simulativa of the Universita' del Piemonte Orientale and described in detail in Ortona, 2005, provided the value 0.696 for p (with 0.402 standard deviation). However, *each* participant to the experiment provided his/her value; given these values, it was tedious but simple to apply the choice method described in this paper to the ten systems above *and to each participant*. It is not inappropriate to state that participants *voted* their

---

[6]For the proof see (Fragnelli et al., 2005).

[7]A Condorcet cycle may result only by chance, and may be ruled out simply by adding a further figure while rounding the results - or by tossing a coin.

Table 1.   A simulation of an Italian-like case.

	System	r	g	share of seats of the governing coalition	parties of the governing coalition
1	Borda	0.66	0.275	0.55	2
2	Run-off plurality	0.66	0.300	0.60	2
3	Plurality	0.74	0.233	0.70	3
4	Mixed-sc. (a)	0.85	0.207	0.61	3
5	Mixed (a)	0.82	0.207	0.62	3
6	Prop. (1 district)	1.00	0.104	0.52	5
7	Threshold Prop. (b)	0.87	0.170	0.51	3
8	Condorcet	0.70	0.295	0.59	2
9	Prop. Hare (c)	0.92	0.135	0.54	4
10	Prop. Imperiali (c)	0.88	0.087	0.52	6
11	Prop. Sainte-Lague (c)	0.94	0.135	0.54	4
12	Prop. D'Hondt (c)	0.84	0.180	0.54	3
13	STV N.B. (c)	0.94	0.106	0.53	5
14	STV Droop (c)	0.95	0.108	0.54	5
15	STV Hare (c)	0.91	0.108	0.54	5
16	Prop. Hare (d)	0.99	0.106	0.53	5
17	Prop. Imperiali (d)	0.99	0.106	0.53	5
18	Prop. Sainte-Lague (d)	0.98	0.108	0.54	5
19	Prop. D'Hondt (d)	0.96	0.104	0.52	5
20	Mixed-sc (d)	0.91	0.177	0.53	3
21	Mixed (d)	0.87	0.190	0.57	3
22	Threshold prop. (b, d)	0.96	0.106	0.53	5

(a) 25 seats assigned through one-district proportionality, 75 through plurality.
'sc' (after the Italian word 'scorporo') means that votes used for the proportional share are
    not considered for the assignment of the plurality seats.
(b) Threshold 5%.
(c) Ten ten-seat districts.
(d) Five twenty-seat districts. The program ran out of memory for STV.
Simulations were performed with 100 seats.
• prop. = pure proportionality
• STV = single transferable vote

preferred electoral system. Condorcet got 46 votes, pure proportionality 17, mixed (5 districts) 12, and mixed (ten districts) 5. The Condorcet winner is Condorcet; a result hardly unexpected for a theorist, but not that easy to detect from data, as Condorcet ranks second in g but only second to last in r.

4.   The Role of Power

We think that representativeness and governability should take into account more than the distribution of seats w.r.t. the distribution of votes. Mathematics

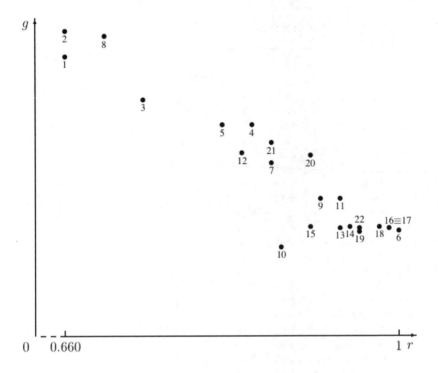

Fig. 1. The voting systems of Table 1 in the $r - g$ space.

offers a lot of distances or norms in order to measure the distance of the distribution of voters, v_i and the distribution of seats according a system h, s_i^h; among them the most widely used are:

- Norm 1: $d_1^h = \sum_{i \in N} |v_i - s_i^h|$

- Norm 2: $d_2^h = \sqrt{\sum_{i \in N} (v_i - s_i^h)^2}$

- Norm ∞: $d_\infty^h = \max_{i \in N} |v_i - s_i^h|$

where N is the set of parties.

This approach may be largely far from our needs, as shown in the following example.

EXAMPLE 2 [8] *Suppose that there are four parties P_A, P_B, P_C and P_D; the preferences of the voters are respectively 40, 25, 20 and 15 per cent and the*

---

[8] Taken from (Fragnelli et al., 2005).

majority quota is 50 per cent; suppose also that the parliament consists of 4 seats and that two voting systems generate the two partitions of seats (2,1,1,0) and (1,1,1,1). We start by computing the distances of the two partitions from the distribution of voters:

	(2,1,1,0)	(1,1,1,1)
d_1	0.4	0.4
d_2	$0.01 \sqrt{350}$	$0.01 \sqrt{350}$
d_∞	0.15	0.15

The two voting systems seem to be equivalent.   ◇

In order to avoid these unlikely situations we can relate the indices not directly to the distributions of votes and seats, but to the *power* of the parties.

The elusive notion of power has a lot to do with the choice of the electoral system; and both with governability and with representativeness. If we stick to the microcosm notion of representativeness, we should want a distribution of power similar to that of preferences, while the governability is normally supposed to be enhanced if the power is highly concentrated. To find out the 'right' distribution of power is a formidable task, and we will not deal with it. More modestly, we argue that in order to tackle that problem it is necessary to be able to compare the distribution of power with that of preferences; and again simulation is highly useful, for the same reasons that we discussed above - the non-availability of reliable real world data.

So, the new problem we have to face is to determine the distribution of the power of the parties.

Game theory is a natural habitat for the problem of evaluating the power of the parties in a voting situation. Since the pivotal paper of Shapley (1953) different indices were introduced, with the aim of assigning to each agent a number that represents his/her relevance in a multiagent situation. It may be useful to recall some basic notions. A *cooperative game with transferable utility (TU-game)* is a pair $G = (N, v)$, where N is the set of players (the agents) and v is the characteristic function that assigns to each subset of players $S \subseteq N$, called *coalition*, a real number that can be considered as its worth independently from the behavior of the other players. A game is said to be *simple* if $v(S) \in \{0, 1\}$, i.e. the worth of a coalition may be only 0 or 1; a game is said to be *monotonic* if $S \subset T$ implies $v(S) \leq v(T)$, i.e. if a coalition is enlarged then its worth cannot decrease.

In particular we are interested in the *weighted majority games*, simple monotonic games that are widely used in voting situations. Suppose that each player $i \in N$ is associated with a non negative real number, the *weight* w_i, and suppose that if some players join to form a coalition S the weight of the coalition is the sum of the weights of the players, i.e. the weights are additive; if the weight of a coalition is strictly larger than a given positive real number q, the so-called *quota*, the coalition is said to be *winning*, and it is said to be *losing* otherwise. Formally we define the characteristic function w of a weighted majority game as:

$$w(S) = \begin{cases} 1 & \text{if } \sum_{i \in S} w_i > q \\ 0 & \text{if } \sum_{i \in S} w_i \leq q \end{cases} \qquad \forall \, S \subseteq N$$

Usually such a situation is summarized by the $(n+1)$-upla $(q; w_1, ..., w_n)$.

As a consequence we can say that if $v(S) = 1$ then S is a winning coalition and if $v(S) = 0$ then S is a losing coalition. A winning coalition is called *minimal* if all its subcoalitions are losing.

The weighted majority games associated to the distributions of voters and of seats, according to a given electoral system, allow us evaluating the importance of each party with respect to a suitable power index. Game theory dealt with this problem from the beginning of its history. Many different power indices were proposed, each of them emphasizing different properties of the underlying situation. In this paper we consider the Shapley-Shubik index, the normalized Banzhaf-Coleman index, the Deegan-Packel index and the Holler (or Public goods) index.

The *Shapley-Shubik index* (Shapley and Shubik, 1954), ϕ, is the natural extension of the Shapley value (Shapley, 1953) to simple games. Let Π be the set of all the permutations of the players and for each $\pi \in \Pi$ let $P(i, \pi)$ be the set of players that precede player i in π; the Shapley value is the average marginal contribution of each player w.r.t. the possible permutations [9]:

$$\phi_i = \frac{1}{|N|!} \sum_{\pi \in \Pi} [v(P(i, \pi) \cup \{i\}) - v(P(i, \pi))] \qquad \forall \, i \in N$$

The *normalized Banzhaf-Coleman index* (Banzhaf, 1965 and Coleman, 1971), β, is similar to the Shapley-Shubik index, but it considers the marginal contributions of a player to all possible coalitions, without considering the order of the players. Let us introduce $\beta_i^* = \frac{1}{2^{|N|-1}} \sum_{S \subseteq N, S \ni i} [v(S) - v(S \setminus \{i\})], i \in N$. By normalization we get:

$$\beta_i = \frac{\beta_i^*}{\sum_{j \in N} \beta_j^*} \qquad \forall \, i \in N$$

The *Deegan-Packel index* (Deegan and Packel, 1978), δ, considers only the minimal winning coalitions; the power is firstly equally divided among minimal winning coalitions and then the power of each is equally divided among

---

[9] We denote by $|A|$ the cardinality of the set A

its members:

$$\delta_i = \sum_{S_k \in W, S_k \ni i} \frac{1}{|W|} \frac{1}{|S_k|} \qquad \forall\, i \in N$$

The *Holler index*, or *Public Goods index* (Holler, 1982 and Holler and Packel, 1983), H, considers the number of minimal winning coalitions which player i belongs to, $c_i, i \in N$; then by normalization we get:

$$H_i = \frac{c_i}{\sum_{j \in N} c_j} \qquad \forall\, i \in N$$

The different indices take into account various aspects of the coalition formation process, so that the power of a given party may assume different values. In particular the power could be concentrated in few large parties or spread on many of them.

EXAMPLE 3 *Referring to Example 2, we can define the majority games $w(v)$ on voters, $w(s^1)$ on the first parliament and $w(s^2)$ on the second parliament*

game	1	2	3	4	12	13	14	23	24	34	123	124	134	234	N
$w(v)$	0	0	0	0	1	1	1	0	0	0	1	1	1	1	1
$w(s^1)$	0	0	0	0	1	1	0	0	0	0	1	1	1	0	1
$w(s^2)$	0	0	0	0	0	0	0	0	0	0	1	1	1	1	1

whose corresponding indices are:

game	ϕ	β	δ	H
$w(v)$	$\left(\frac{1}{2},\frac{1}{6},\frac{1}{6},\frac{1}{6}\right)$	$\left(\frac{1}{2},\frac{1}{6},\frac{1}{6},\frac{1}{6}\right)$	$\left(\frac{9}{24},\frac{5}{24},\frac{5}{24},\frac{5}{24}\right)$	$\left(\frac{1}{3},\frac{2}{9},\frac{2}{9},\frac{2}{9}\right)$
$w(s^1)$	$\left(\frac{2}{3},\frac{1}{6},\frac{1}{6},0\right)$	$\left(\frac{3}{5},\frac{1}{5},\frac{1}{5},0\right)$	$\left(\frac{1}{2},\frac{1}{4},\frac{1}{4},0\right)$	$\left(\frac{1}{2},\frac{1}{4},\frac{1}{4},0\right)$
$w(s^2)$	$\left(\frac{1}{4},\frac{1}{4},\frac{1}{4},\frac{1}{4}\right)$	$\left(\frac{1}{4},\frac{1}{4},\frac{1}{4},\frac{1}{4}\right)$	$\left(\frac{1}{4},\frac{1}{4},\frac{1}{4},\frac{1}{4}\right)$	$\left(\frac{1}{4},\frac{1}{4},\frac{1}{4},\frac{1}{4}\right)$

Finally, for each index, we compute the distances between the power w.r.t. the voters and to each parliament:

	(2,1,1,0)				(1,1,1,1)			
	ϕ	β	δ	H	ϕ	β	δ	H
d_1	**0.333**	**0.333**	0.417	0.444	0.500	0.500	**0.250**	**0.167**
d_2	0.236	0.200	0.250	0.281	0.289	0.289	**0.144**	**0.096**
d_∞	**0.167**	**0.167**	0.222	0.208	0.250	0.250	**0.125**	**0.083**

where numbers in bold indicate the 'best' voting system according to each index.
The distances on the power indices allow us to distinguish the two systems. In particular the indices of Shapley-Shubik and Banzhaf-Coleman favor the first parliament, while the second parliament is preferable according to the indices of Deegan-Packel and Holler. ◇

Another measure may be obtained referring to the distribution of voters, v, to the assignment of seats according to an electoral system h, s^h, and to the power of the parties related to the votes and to the seats, φ and φ^h respectively (see Gambarelli and Biella, 1992). The resulting distance Δ is:

$$\Delta = \max_{i \in N} \left| |v_i - s_i^h| - |\varphi - \varphi^h| \right|$$

EXAMPLE 4 *Referring again to Example 2, we obtain the following distances:*

Voting system	$\varphi = \phi$	$\varphi = \beta$	$\varphi = \delta$	$\varphi = H$
s^1	**0.067**	**0.033**	**0.058**	**0.072**
s^2	0.100	0.100	**0.058**	**0.072**

where again numbers in bold indicate the 'best' voting system according to each index.
Also using this measure the Shapley-Shubik and Banzhaf-Coleman favor the first parliament, while the two parliaments are equivalent according to the indices of Deegan-Packel and Holler. ◇

REMARK 5 *The first two indices and the last two have similar behavior; this depends on the matter that the first two take into account all the coalitions, while the last two consider only minimal winning coalitions.*

The main conclusions of this section are the following. First, the indication of the example - were it for real - would be precious. Yet the starting point are, by necessity, the data on votes. If votes are those actually cast in a, say, plurality election, they are useless to compare the distribution of power with that of *preferences*. Arguably, the distribution of votes may be assumed as a proxy to that of preferences only in proportional systems with large districts (and, we must add, with low running costs). The same conclusions of Section 3 apply. Real data cannot provide useful information; the simulation does. To accumulate experimental (i.e. simulative) evidence would probably provide relevant suggestions for real world analysis and policing.

Second, in our opinion, a better definition of both representativeness and governability should rest on the notion of power. Game theory is a very useful tool for this aim. In particular the index of concentration (Gini, 1914) applied to the distribution of power may be exploited to define a representativeness index, while for the governability index two approaches seem promising: the coalitional value (Owen, 1977), that takes into account the role of the *a priori* agreements of the parties and the propensity to disruption (Gately, 1974) that measures the relative gain of the players when leaving the grand coalition.

5.   Concluding Remarks

We argued that the simulation approach to the evaluation of electoral systems is very powerful. To add evidence, we suggest a (very partial) list of problems that could profitably be tackled this way. What is the difference *in results* between Borda and Condorcet? When do pure proportionality and single transferable vote provide analogous results? What is the actual effect of district magnitude on proportionality? How do indices of proportionality perform? Are Condorcet cycles really a problem?

It is not difficult to add others, so we will not pursue this point further. Instead, we argue that experimental results may be improved if the simulation programs are further elaborated. We suggest that main methodological improvements should regard the possibility of including and managing survey data, the addition of further indices, mostly but not only with reference to power issues, and obviously the addition of further electoral systems. However, to our opinion the main methodological challenge is the addition of new evaluation *dimensions*, and consequently indicators. Obviously, this requires that they may be quantified, and consensus on what we desire about.

To conclude, simulation is very useful to analyze the performance of the electoral systems including random elements, e.g. the absence of some members of the parliament in a voting session, or to study the possibility of manipulating the elections, e.g. via merging or splitting of the parties in order to profit of suitable features of the system.

Appendix A - The Simulation Program

Given the utility and the versatility of the simulative approach for the analysis of electoral systems, it is quite surprising that it is so little employed in the political science literature. A survey is in (Fragnelli et al., 2005); there are some, but not that many, suggestive case studies, but very few papers address the matter we are dealing with here, to compare electoral systems, after some pioneering papers (see Mueller, 1989; Merrill, 1984 and Merrill, 1985). (Gambarelli and Biella, 1992) analyze the effect in Italy of a change to a number of electoral systems, and (Christensen, 2003), compares six majoritarian systems, but without reference to a Parliament. Consequently, it is not surprising that the simulation programs so far available (like those developed by Accuratedemocracy, www.accuratedemocracy.org) are of limited use for purposes of the kind suggested here.

The simulations produced in this paper have been carried out with a specific program, ALEX4; its number refers to the version currently in use. ALEX4 is written in Java, and it is the heir of a program originally written in Visual Basic, $g\&r$ (for Governability and Representativeness), dating back to 1998 (Ortona, 1998; Trinchero, 1998). ALEX4 is a cosmetic improvement of ALEX3, which is described in detail in (Bissey et al., 2004); hence here we provide only some basic hints. All the versions of ALEX have been written by Marie-Edith Bissey at the Universita' del Piemonte Orientale.

The user is requested to provide some basic inputs, namely, the size of the Parliament, the number of voters, the number of parties (i.e. the number of candidates in every constituency for non-proportional systems), the share of votes of the parties, the concentration of the parties across the constituencies, the probability that second and further preferred parties are the closest to the first, to the second etc., the probability that third and further preferred parties are the closest to the second, to the third etc., and the probability that second and further preferred candidates are the closest to the first, the second, etc. The first two probabilities are employed to generate a complete set of preferences for parties, for each voter; the last one to generate a complete set of preferences for candidates, to be employed for single transferable vote. The program produces the Parliaments for (up to now) sixteen systems, namely one-district proportionality; one-district threshold proportionality [10]; Hare, D'Hondt, Imperiali and Sainte-Lague multi-district proportionality; N.B., Droop and Hare multi-district single transferable vote; two mixed-member systems [11]; plurality; run-off majority; Condorcet; Borda; and VAP, a suggested new system described in detail in (Ortona, 2004). Approval voting is not included (but it will be in further versions) because previous experiments (with $g\&r$) indicated that it is commonly dominated by other systems. Finally, the program computes the index of representativeness and the index of governability (the user is requested to define the governing coalition). Both indices are described in Appendix B.

Appendix B - The Indices Employed

Index of Representativeness, r

An index of representativeness suitable to compare electoral systems cannot be based on the difference between the share of votes and that of seats, albeit all the indices of *proportionality* commonly employed, like Gallagher's [12], are constructed this way. The obvious and compelling reason is that the voting behavior is affected by the electoral system itself. Instead, our index is based on the difference between *votes cast in a nation-wide proportional district and*

---

[10]The threshold may be fixed by the user.
[11]With and without the exclusion of votes employed in the plurality election from the proportional election. The share of seats assigned through proportionality may be fixed by the user.
[12]Introduced by (Gallagher, 1991).

seats assigned by a given electoral system. The formula is:

$$r_h = 1 - \frac{\sum_{i \in N} |S_i^h - S_i^{PP}|}{\sum_{i \in N} |S_i^u - S_i^{PP}|}$$

where N is the set of parties, S_i^h is the number of seats of party i with system h, S_i^{PP} is the number of seats of party i with the perfect proportional system and S_i^u is the total number of seats for the relative majority party under system h and it is 0 otherwise.

The index reads as follows. For the sum at the numerator, we assume that the representativeness R is maximal under perfect proportionality rule (PP). Hence the loss of representativeness incurred by party i is the (absolute) difference between the seats it would get under PP and those actually obtained. Summing this loss across all the parties we obtain the total loss of R. The sum at the denominator is introduced to normalize this value. It is the maximum possible loss of R. This maximum is obtained when 'winner takes all' in a very strict sense, that is when the relative majority party, according to the selected system, takes all the seats instead of just its quota. The ratio of the sums is a loss of representativeness index, normalized in the range 0-1; subtracting it from 1 we transform it into a representativeness index.

EXAMPLE 6 *Suppose three parties, L, C and R, in a parliament of 100 seats. Under PP they obtain 49, 31 and 20 seats respectively, under majority (M) 90, 10 and 0, and under some other system (S) 30, 55 and 15. So* $r_M = 1 - \frac{41+21+20}{51+31+20} = 0.196$ *and* $r_S = 1 - \frac{19+24+5}{49+45+20} = 0.579$ *(obviously* $r_{PP} = 1 - \frac{0}{51+31+20} = 1)$ ◇

As this index is not that easy to grasp, in the example described in Section 3 above we employed a simpler one, which is 1 minus the ratio between the total number of seats assigned in excess to the proportional share and the total number of seats. In the previous example the value of this second index is 0.59 for M and 0.89 for S (and 1 for PP). For more realistic cases, however, the two indices are strongly correlated; for data of Section 3 the correlation index is 0.963.

Index of Governability, g

According to the mainstream doctrine, governability is inversely related to the number of parties that take part in the governing majority. Our index is based on this assumption. It depends on the number of parties of the governing coalition that may destroy the majority if they withdraw, m, and on the share of seats of the majority, f. m is more important, so we add (lexicographically) the f-component to the m-component. Hence the index is made by the sum of two terms, the first related to m, g_m, and the second related to f, g_f. Thus,

$g = g_m + g_f$. The range of the second term is the difference between successive values of the first: the term in m defines a lower and an upper bound, and the term in f specifies the value of the index between them.

The range defined for g_m is simply $\frac{1}{m}$ (upper bound) and $\frac{1}{m+1}$ (lower bound). For instance, if the government is supported by just one party, g is in between 0.5 and 1; if it supported by two parties, then g is in between 0.333 and 0.5, and so on. The number of seats of the majority coalition specifies the value of g in the given range. The amount g_f to be added to the lower bound depends from the lead of the majority coalition, according to the proportion $\frac{g_f}{\frac{1}{m} - \frac{1}{m+1}} = \frac{f - \frac{T}{2}}{T - \frac{T}{2}}$. In sum, the formula for g is:

$$g = g_m + g_f = \frac{1}{m+1} + \frac{1}{m(m+1)} \frac{f - \frac{T}{2}}{\frac{T}{2}}$$

For instance, if there are 100 seats and the governing majority is made up of one party with 59 members, we have $g_f = \frac{9}{50}\frac{1}{2} = 0.09$. This value must be added to 0.5, to give $g = 0.59$.

The maximum value of g is 1, when a party has all the seats; the lowest tends to zero as the number of parties increases, thus justifying the claim that the range of g is in between 0 and 1.

Again, in Section 3 we employed a simpler index, based on the same theoretical assumptions, i.e. the ratio between the share of seats and the number of parties of the governing coalition. In the example, the value of this index[13] is again 0.59; and this index is strongly correlated with the previous one; for data of Section 3 the correlation index is 0.994.

Appendix C - A Short Description of the Electoral Systems

This appendix is taken from the *readme* file of ALEX4 package. Many systems allow for variants; the definitions provided here refer to those adopted in ALEX4. For a description of how the systems are implemented, see the *Final Note* of this paper, and the *readme* file quoted. For an easy-to-read, more detailed description of the systems, see (Farrell, 2001).

- *Plurality* In each district, the winner is the candidate with most votes.

---

[13] There is a reason for dissatisfaction with indices of governability based on the number of parties, which to our opinion is why this kind of indices perform quite poorly when applied to real cases, and more generally why the governability is not that greater in majoritarian systems (see Lijphart, 1999). The reason is that a party may be and may be not a single subject. At one extreme it is, but at the other it is a set of independent decision-makers. ALEX5, the next version of the program ,will take into account this feature through the addition of a suitable parameter.

- *One-district Proportionality* The seats in the Parliament are distributed according to the shares of votes in the population, rounded to the closer integer.

- *Threshold Proportionality* All the parties who have a share of votes in the population smaller than the established threshold are excluded from the Parliament. The seats are distributed proportionally among the remaining parties.

- *Run-off Plurality* In each district all parties but the two with the most votes are excluded. The second round is carried out with these two parties only and the one with the most votes wins. If after the first round the first party has at least 50% of the votes, it wins the seat without the need of a second round.

- *Mixed* Part of the parliament is elected with the Plurality System, and the rest is elected using the Proportional System.

- *Mixed with 'Scorporo'* As for the previous system, but the votes used to elect the Plurality share are lost for the Proportional share.

- *Borda count* This system uses the electors' complete preference ordering. Each elector gives points to each party, from 0 for the most preferred party to $N - 1$ for the least preferred party, where N is the total number of parties. In each district, the winner is the party with fewer points.

- *Condorcet winner* In each district, the Condorcet winner is the party that beats all the others when taken in pairs.

- *Multi-district Proportionality* The method is the same as in one-district proportionality. In this case, however, the rounding procedure is relevant. We employed four: D'Hondt, Hare, Imperiali and Sainte-Lague.

- *Single Transferable Vote* The seats for each party, in each plurinominal district, are assigned according to a quota value. If some seats are not assigned with this method, the votes unused by the elected candidates are transferred to the next candidates in the elector's preference ordering, and the candidates with the highest number of votes (obtained + transferred) are elected. The quota value may be computed according to three different procedures: N.B., Droop, and Hare.

Final Note

If you are interested, you may download and use the simulation program ALEX4. There will be no charge, but you will be asked to observe some gentleperson-agreement conditions - basically, no liability for possible mistakes and quotation of the source of the program. Please contact Guido Ortona for further details or for downloading instructions.

Acknowledgments

The authors gratefully acknowledge the suggestions of the referees and the financial support of Italian MIUR, grant PRIN2005.

References

Banzhaf, J.F. (1965). "Weighted Voting doesn't Work: A Mathematical Analysis," Rutgers Law Review, 19, 317-343.

Bissey, M.E., M. Carini and G. Ortona (2004). "ALEX3: A Simulation Program to Compare Electoral Systems," Journal of Artificial Societies and Social Simulation 7 (electronic journal at http://jasss.soc.surrey.ac.uk).

Bowler, S., D.M. Farrell and R.T. Pettitt (2005). "Expert Opinion on Electoral Systems: So Which Electoral System is 'best'?," Journal of Elections, Public Opinion and Parties, 15, 3-19.

Christensen, N.B. (2003). Evaluating Voting Procedures Using Different Criteria and Computer Simulations, Conference of the European Public Choice Society, Aarhus.

Coleman, J.S. (1971). "Control of Collectivities and the Power of a Collectivity to Act," in B. Lieberman (Ed.), Social Choice, London: Gordon and Breach, 269-300.

Deegan, J., and E.W. Packel (1978). "A New Index of Power for Simple n-person Games," International Journal of Game Theory, 7, 113-123.

Farrell, D.M. (2001). Electoral Systems - A Comparative Introduction, Basingstoke: Palgrave.

Fragnelli, V., G. Monella and G. Ortona (2005). "A Simulative Approach for Evaluating Electoral Systems," Homo Oeconomicus, 22, 524-549.

Gallagher, M. (1991). "Proportionality, Disproportionality and Electoral Systems," Electoral Studies, 10, 33-51.

Gambarelli, G., and R. Biella (1992). "Sistemi elettorali," Il Politico, 164, 557-588.

Gately, D. (1974). "Sharing the Gains from Regional Cooperation: A Game Theoretic Application to Planning Investment in Electric Power," International Economic Review, 15, 195-208

Gini, C. (1914). "Sulla misura della concentrazione e della variabilita' dei caratteri" Atti del Regio Istituto Veneto di Scienze, Lettere ed Arti, 73, 1203-1248.

Holler, M.J. (1982). "Forming Coalitions and Measuring Voting Power," Political Studies, 30, 262-271.

Holler, M.J., and E.W. Packel (1983). "Power, Luck and the Right Index," Zeitschrift für Nationalökonomie (Journal of Economics), 43, 21-29.

Lijphart A. (1999). Patterns of Democracy. Government Forms and Performance in Thirty-Six Countries, Yale: Yale University Press.

Merrill, S. III (1984). "A comparison of Efficiency of Multicandidate Electoral Systems," American Journal of Political Science, 28, 49-74.

Merrill, S. III (1985). "A Statistical Model for Condorcet Efficiency based on Simulation under Spatial Model Assumptions," Public Choice, 47, 389-403.

Mueller, D.C. (1989). Public Choice III, Cambridge: Cambridge University Press.

Ortona, G. (1998). "Come funzionano i sistemi elettorali: un confronto sperimentale," Stato e Mercato, 54: 469-486.

Ortona, G. (2002). "Choosing the electoral system: why not simply the best one?," Department POLIS, Universita' del Piemonte Orientale, Working Paper 32.

Ortona, G. (2004). "Un nuovo sistema elettorale a due stadi e una sua valutazione mediante simulazione," Polena, 2, 49-66.

Ortona, G. (2006). "Voting on the Electoral System: an Experiment," Polena, forthcoming.

Owen, G. (1977). "Values of Games with a Priori Unions," Lecture Notes in Economic and Mathematical Systems, 141, 76-88.

Shapley, L. S. (1953). "A Value for n-Person Games," in Contributions to the Theory of Games, Vol II (Annals of Mathematics Studies 28) (Kuhn HW, Tucker AW eds.), Princeton: Princeton University Press, 307-317.

Shapley, L.S., and M. Shubik (1954). "A Method for Evaluating the Distribution of Power in a Committee System," American Political Science Review 48, 787-792.

Trinchero, R. (1998). "Sistemi di voto e preferenze individuali: un'analisi comparata con l'uso della simulazione al calcolatore," Working Paper 9802, Department of Economics, Universita' di Torino.

How to Elect a Representative Committee Using Approval Balloting

D. Marc Kilgour[1], Steven J. Brams[2], M. Remzi Sanver[3]

[1]Department of Mathematics, Wilfrid Laurier University

[2]Department of Politics, New York University

[3]Department of Economics, Istanbul Bilgi University

Abstract   Approval balloting is applied to the problem of electing a representative committee. We demonstrate several procedures for determining a committee based on approval ballots, paying particular attention to weighting methods that can reduce the influence of voters with extreme views. We show that a general class of voting systems based on approval ballots can be implemented through analysis of appropriate tables. A by-product of this procedure is a clarification of the complexity of these systems.

Keywords: Approval balloting, committee election, complexity.

Introduction

Approval voting is a well-known voting procedure applicable to single-winner elections. Voters approve of as many candidates as they like, and the candidate with the most approvals wins (Brams and Fishburn, 1978, Brams and Fishburn, 1983, Brams and Fishburn, 2005). But this method of aggregating approval votes is not the only one possible, as Merrill and Nagel (1987) argue. It is therefore useful to distinguish between *approval balloting* (each voter submits a ballot that identifies which candidates are approved) and *approval voting* (the method, indicated above, by which approval ballots are tallied to determine the winner).

In this paper, we discuss how approval ballots can be used to select a committee—a subset of the candidates—that represents, in some sense, all voters. In such an election, voters would be instructed to indicate on their approval ballots the subsets of candidates that best represent them.

Our procedures for identifying a most representative subset of the set of candidates capitalize on the fact that each ballot also specifies a subset of this set. Of course, different voters will typically specify different subsets. We view the problem of identifying the most representative committee as that of identifying the subset that is "closest" to the collection of subsets specified by the voters. Our procedures can be adapted to reflect restrictions on the size or composition of the committee to be elected.

Based on an appropriate measure of distance, we discuss two criteria of closeness to the collection of subsets specified by the voters, *minimax* (a representative subset should minimize the maximum distance to the subsets of all voters) and *minisum* (a representative subset should minimize the sum of distances or, equivalently, the average distance to the subsets of all voters). Elsewhere we offer a broader discussion of criteria of fairness in electing committees (Brams, Kilgour, and Sanver, 2005).

1.   Terminology and Notation

There are $n > 1$ voters. The set of $k > 1$ candidates is denoted $C = \{1, 2, \ldots, k\}$. We represent a subset of the candidates $S \subseteq C$ by a (row) vector $p = (p_1, p_2, \ldots, p_k)$, where $p_h = 1$ if $h \in S$ and $p_h = 0$ otherwise. (Usually we will write subsets in vector notation without punctuation—for example, 1001101 designates the subset comprising candidates 1, 4, 5, and 7.)

The n voters' ballots are $p^1, p^2, \ldots, p^n$, which we write as rows of a 0-1 matrix, P, called the *ballot data matrix*. Note that P has n rows (corresponding to the voters) and k columns (corresponding to the candidates); the entry in row i and column h of P, p_h^i, is 1 if voter i approves of candidate h and 0 otherwise.

EXAMPLE 1 *There are $n = 3$ voters and $k = 3$ candidates. Voter 1's ballot is 100 (i.e. voter 1 approves of candidate 1 only), whereas voter 2's ballot is 110, and voter 3's ballot is 101. The ballot data matrix is*

$$P = \begin{pmatrix} 1 & 0 & 0 \\ 1 & 1 & 0 \\ 1 & 0 & 1 \end{pmatrix}$$

The ballot data matrix records the subset of candidates approved by each voter. Because we wish to construct anonymous voting systems, we need not maintain a record of which voter approved of a particular subset. Moreover, as the number of voters increases (and the number of candidates becomes relatively small), it is increasingly likely that some voters will cast identical ballots—that is, approve of exactly the same subset. If so, the ballot data matrix will contain many identical rows. To simplify the data, we record only the distinct ballots (in any order), and the number of times each is chosen.

More specifically, we associate with the voted-for subsets, $q^1, q^2, \ldots, q^\ell$, corresponding counts $m_1, m_2, \ldots, m_\ell$, indicating that $m_j > 0$ different voters approve of exactly the subset q^j. It follows that $\sum_{j=1}^{\ell} m_j = n$ and $q^j \neq q^{j'}$ whenever $j \neq j'$. As before, we write $q^1, q^2, \ldots, q^\ell$ as rows of a 0-1 matrix

Q, called the *compressed ballot data matrix*. Note that Q has ℓ rows and k columns, and the entry in row j and column h of Q is denoted q_h^j. Associated with Q is a *count vector*, m, which is a column vector with the count, m_j, as its jth entry. We call (Q, m) *compressed ballot data*.

EXAMPLE 2 *There are $n = 4$ voters and $k = 3$ candidates. Voters 1, 2, and 3 vote as in Example 1; voter 4 votes exactly as voter 3. The compressed ballot data, (Q, m), are*

$$Q = \begin{pmatrix} 1 & 0 & 0 \\ 1 & 1 & 0 \\ 1 & 0 & 1 \end{pmatrix}, \quad m = \begin{pmatrix} 1 \\ 1 \\ 2 \end{pmatrix}$$

To measure the distance between two subsets of C (possible committees), p and q, we will use the *Hamming distance*, $d(p, q) = \sum_{h=1}^{k} |p_h - q_h|$. Thus, $d(p, q)$ equals the number of components of p and q that are different, or the number of candidates who are in one of p or q but not the other. Note that for any subsets p and q of C, $0 \le d(p, q) \le k$ and, of course, $d(p, q) = 0$ iff $p = q$. To illustrate using Examples 1 and 2, $d(100, 110) = 1$ (the two subsets differ only on candidate 2), and $d(110, 101) = 2$ (the subsets differ on candidates 2 and 3).

2. Minisum and Minimax Criteria

We begin by addressing the problem of selecting a representative subset of candidates, p, given a ballot data matrix, P. We assume that there are no restrictions on the subset to be selected.

This problem was considered by Brams, Kilgour and Sanver (2004). One solution they proposed was majority voting (MV), which can be implemented on P by summing each column to obtain the total vote for each candidate. An MV committee is a committee comprising only candidates who receive at least $\frac{n}{2}$ votes.[1] There is always at least one MV committee; in the extreme case in which every candidate receives fewer than $\frac{n}{2}$ approvals, the MV committee contains no members, that is $00 \ldots 0$.

Formally, for ballot data P, the number of votes for candidate h is $n_P(h) = \sum_{i=1}^{n} p_h^i$. Define the conditions $Y_h = Y_h(P)$ and $N_h = N_h(P)$ as follows: Y_h is True if $n_P(h) > \frac{n}{2}$ and False otherwise; N_h is True if $n_P(h) < \frac{n}{2}$ and False

---

[1] We adopt this definition, rather than the standard requirement of more than $\frac{n}{2}$ votes, for technical reasons that will become apparent shortly. Our definition is equivalent to the conventional one whenever n is odd; when n is even, differences are unlikely unless n is small. Our definition implies that the MV committee is not unique (i.e., two or more subsets are tied for winning) if and only if n is even and at least one candidate receives precisely $\frac{n}{2}$ votes. In this case, an MV committee includes all candidates who receive more than $\frac{n}{2}$ votes, plus any subset of the set of candidates who receive exactly $\frac{n}{2}$ votes.

otherwise. Thus, Y_h is true iff candidate h has majority support, and N_h is true iff candidate h is opposed by a majority. The set of all MV committees is then

$$MV(P) = \{p \subseteq C : \forall h = 1, \ldots, k, p_h = 1 \text{ if } Y_h \text{ and } p_h = 0 \text{ if } N_h\}. \quad (1)$$

For instance, in Example 1, the total votes for candidates 1, 2, and 3 (the column sums of P) are $n_P(1) = 3$, $n_P(2) = 1$, and $n_P(3) = 1$, respectively. Because only candidate 1 receives more than $\frac{n}{2} = \frac{3}{2}$ votes, and no candidate receives exactly $\frac{3}{2}$ votes, the unique MV committee is 100 (i.e., it includes only candidate 1).

Brams, Kilgour and Sanver (2004) proved (Proposition 4) that a subset of the candidates, p, is an MV winner if and only if it minimizes $\sum_{i=1}^{n} d(p, p^i)$. Thus, the MV winners are the subsets of candidates that are at minimum total distance (or, equivalently, at a minimum average distance) from the voters' ballots. For this reason, they referred to an MV committee as a *minisum* committee. We define

$$minisum(P) = MV(P).$$

There are two or more committees in $minisum(P)$ whenever at least one candidate receives precisely $\frac{n}{2}$ votes. If so, the total distance from the ballots to a committee containing such a candidate is exactly equal to the total distance from the ballots to the same committee without the candidate, rendering both of these committees members of $minisum(P)$.

As a second approach to finding a representative committee, Brams, Kilgour and Sanver (2004) adapted the unanimity version of the Fallback Bargaining procedure (Brams and Kilgour, 2001). They proposed an iterative procedure that takes place in discrete time, $t = 0, 1, 2, \ldots$. At time t, voter i is modeled as willing to be represented by any subset of candidates in his or her *acceptable set*

$$A_P^i(t) = \{p \in C : d(p^i, p) \leq t\}. \quad (2)$$

For example, at time $t = 0$, voter i's only acceptable subset is p^i, the subset specified on his or her approval ballot. At time $t = 1$, $A^i(t)$ includes p^i and any subset at Hamming distance 1 from p^i —any subset that differs from p^i in exactly one candidate, who might be a member of p^i now excluded, or a non-member of p^i now included. At time $t = 2$, voter i's acceptable set includes all subsets at Hamming distance at most 2 from p^i, and so on. This fallback process continues until there is a subset that is acceptable to all voters.

Formally, the fallback process ends at time t_P^* defined by

$$t_P^* = \min\{t = 0, 1, \ldots : \bigcap_{i=1}^{n} A_P^i(t) \neq \emptyset\}. \quad (3)$$

The fallback (FB) winners are all subsets acceptable to all voters at time t_P^*. Formally,

$$FB(P) = \left\{ p \subseteq C : p \in A_P^i(t_P^*), i = 1, \ldots, n \right\}. \tag{4}$$

For instance, in Example 1, the three voters' acceptable subsets at times 0 and 1 are shown in Table 1.

Table 1.   The fallback process applied to Example 1.

Time (t)	$A^1(t)$	$A^2(t)$	$A^3(t)$
0	{100}	{110}	{101}
1	{100,000,110,101}	{110,010,100,111}	{101,001,111,100}

Observe that the acceptable sets at time $t = 0$ are disjoint, while the acceptable sets at time $t = 1$ have exactly one common member, namely 100. It follows that $t_P^* = 1$ and $FB(P) = \{100\}$. Hence the unique FB committee corresponding to P is 100, according to this iterative procedure in which acceptable sets for each voter become larger and larger over time until there is at least one subset in common.

Based on (Brams and Kilgour, 2001, Theorem 3), Brams, Kilgour and Sanver (2004) showed that a subset, p, is an FB winner if and only if it minimizes $\max_{i=1,\ldots,n} d(p, p^i)$. Thus, the FB winners are the subsets of candidates for which the maximum distance to any voter's ballot is a minimum. For this reason, they referred to an FB committee as a *minimax* committee. We define

$$minimax(P) = FB(P).$$

Brams, Kilgour, and Sanver (2005) next asked whether duplication of ballots could change the set of winning committees. In our more general setting, we consider how the minisum and minimax procedures can be applied to compressed ballot data, (Q, m), as opposed to ballot data, P.

It is immediate that, for compressed ballot data, the number of votes for candidate h is $n_{Q,m}(h) = \sum_{j=1}^{\ell} m_j q_h^j$. Given this emendation, the conditions $Y_h = Y_h(Q, m)$ and $N_h = N_h(Q, m)$ become as follows: Y_h is True if $n_{Q,m}(h) > \frac{n}{2}$ and False otherwise; N_h is True if $n_{Q,m}(h) < \frac{n}{2}$ and False otherwise. Again, Y_h is True iff candidate h has majority support, and N_h is True iff candidate h is opposed by a majority. The set of all minisum committees is then

$$minisum(Q, m) = \{ p \subseteq C : \forall h, p_h = 1 \text{ if } Y_h \text{ and } p_h = 0 \text{ if } N_h \}, \tag{5}$$

essentially unchanged from (1).

When the minimax procedure is applied to ballot data, P, (2) defines the acceptable set of voter i at time t. For compressed ballot data, the acceptable set at time t for a voter who voted for q^j is, analogously,

$$A_{Q,m}^j(t) = \{p \in C : d(q^j, p) \leq t\}. \tag{6}$$

The set of minimax winners can then be determined using

$$t_{Q,m}^* = \min\{t = 0, 1, 2, \ldots : \bigcap_{j=1}^{\ell} A_{Q,m}^j(t) \neq \emptyset\};$$

$$minimax(Q, m) = \left\{p \subseteq C : p \in A_{Q,m}^j(t_{Q,m}^*), j = 1, \ldots, \ell\right\},$$

which is essentially unchanged from (2), (3), and (4).

As Brams, Kilgour and Sanver (2005) noted, the minimax winners are not altered by the duplication of ballots. In our terms, $minimax(Q, m)$ does not depend on the count vector, m, because FB committees reflect only which subsets were voted for, not how many votes each one received. While this property is consistent with some approaches to fairness (the FB committees are the committees for which the worst-represented voter is best represented, which recalls the approach to justice of (Rawls, 1971)), it makes the outcome highly sensitive to outliers and thus not representative of any tendency of voters to cast similar ballots.

Brams, Kilgour and Sanver (2005) suggested that it might be appropriate to revise the minimax procedure so that the rate of increase of the acceptable subset centered at q^j depends inversely on m_j. Applying this variation to compressed ballot data would yield, instead of (6),

$$A_{Q,m}^{j\,\prime}(t) = \left\{p \in C : d(q^j, p) \leq \frac{t}{m_j}\right\}. \tag{7}$$

Then the revised minimax winners would be determined using

$$t_{Q,m}^{*\prime} = \min\{t \in [0, \infty) : \bigcap_{j=1}^{\ell} A_{Q,m}^{j\prime}i(t) \neq \emptyset\};$$

$$minimax'(Q, m) = \left\{p \subseteq C : p \in A_{Q,m}^{j\,\prime}(t_{Q,m}^*), j = 1, \ldots, \ell\right\}.$$

Note that the iterative fallback process still starts at time $t = 0$, but it now takes place in continuous (rather than discrete) time. (Because the m_j are integers, the threshold times $t_{Q,m}^{*\prime}$ are always rational, but we make no use of this simplification.) Note that the minimax committees produced by this adjusted procedure reflect ballot duplication: The more voters who vote for q^j, the slower the acceptable set centered at q^j grows.

Let us illustrate our results so far, taking P as in Example 1 and (Q, m) as in Example 2. We noted earlier that $minisum(P) = \{100\}$. For (Q, m), $n_{Q,m}(1) = 4$, $n_{Q,m}(2) = 1$, $n_{Q,m}(3) = 2$, and $\frac{n}{2} = 2$, so $minisum(Q, m) = \{100, 101\}$. In other words, both 100 and 101 are minisum committees for (Q, m).

We determined earlier that $minimax(P) = \{100\}$. An identical calculation shows that $minimax(Q, m) = \{100\}$. But the action of the revised procedure, which takes into account duplicate ballots, is different, as shown in Table 2.

Table 2.   The revised fallback process applied to Example 2.

Time (t)	$A^{1\prime}(t)$	$A^{2\prime}(t)$	$A^{3\prime}(t)$
0	$\{100\}$	$\{110\}$	$\{101\}$
1	$\{100, 000, 110, 101\}$	$\{110, 010, 100, 111\}$	$\{101\}$
2	$\{100,000,110,101,$ $010,001,111\}$	$\{110,010,100,111,$ $000,011,101\}$	$\{101,001,111,100\}$

Note that at time $t = 1$ the three acceptable sets have no common member, as they did in the previous table (namely, outcome 100). But at time $t = 2$, there are three common members; formally, $t_{Q,m}^{*\prime} = 2$ and $minimax'(Q, m) = \{100, 101, 111\}$.

3.   Weighted Distances

The main objective of this paper is to show how different weightings—which take into account, for example, similarities among ballots—can affect the determination of representative committees. The first observation is simple: If we write $A_j = A_{Q,m}^{j\,\prime}$, then (7) is equivalent to

$$A_j(t) = \{p \in C : w_j d(q^j, p) \le t\}, \tag{8}$$

provided $w_j = m_j$. Thus, whether a particular subset belongs to the acceptable subset centered at q_j can be understood to depend on weighted distances.

It follows from (8) that increasing the weight of q_j tends to draw the winning subset closer to q_j: At any particular time, t, fewer other subsets are acceptable to a voter who supported q^j. One weighting of interest, the *count weight*, is based on the count vector; the weight of a voted-for subset, q^j, is $w_j = m_j$, the number of voters who voted for q^j, that we described earlier.

Another weighting, which we will argue is useful for both minimax and minisum procedures, is *proximity weight*, in which the weight of q^j reflects the extent to which a q^j voter is similar to, or different from, other voters. Specifically,

$$w_j = \frac{m_j}{\sum_{r=1}^{\ell} m_r d(q^j, q^r)}, \tag{9}$$

so that the weight of q^j is proportional to m_j, the number of voters who voted for q^j. The denominator of the fraction in (9) is the sum of the Hamming distances from q^j to the subsets approved by all voters. (Of course, $d(q^j, q^j) = 0$, so the distance from q^j to itself does not contribute to this sum.) Thus w_j, the weight of q^j, is small when few voters approve of exactly q^j or any subset similar to it. By giving them less weight, proximity weighting tends to make the outcome less sensitive to the views of "extreme" voters—that is, voters whose ballots differ substantially from those of most other voters.

To generalize our analysis, consider compressed ballot data (Q, m), and assume that a positive weight, w_j, has been assigned to each q^j. (There is no requirement other than that all weights be positive; in particular, no normalization is assumed—the weights need not sum to 1.) We define the weighted vote for candidate $h = 1, \ldots, k$ to be

$$n_w(h) = \sum_{j=1}^{\ell} w_j q_h^j. \tag{10}$$

A natural threshold of weighted votes is $S = \frac{1}{2} \sum_{j=1}^{\ell} w_j$, or half the total weight. Thus, a weighted-majority rule or MV_w committee is a committee that includes all candidates whose weighted vote is greater than S and no candidates whose weighted vote is less than S.[2] We will show that the MV_w committees are precisely the minisum committees in a weighted-distance context.

We think of $w_j d(p, q^j)$ as the weighted distance between a voted-for subset q^j and a possible committee $p \in C$. Then (8) defines the subsets acceptable to a q^j voter at time t. The unanimity version of the fallback bargaining procedure can be implemented using (8) and

$$t^* = \min\{t \in [0, \infty) : \bigcap_{j=1}^{\ell} A_j(t) \neq \emptyset\}; \tag{11}$$

$$FB_w = \{p \subseteq C : p \in A_j(t^*), j = 1, \ldots, \ell\}. \tag{12}$$

---

[2] We follow the same convention as discussed in the previous footnote. In particular, a candidate whose weighted vote is exactly S may or may not be included in an MV_w committee.

The interpretation is as before: The time t^* is the earliest moment that at least one subset is acceptable to all voters; FB_w is the set of all subsets that are acceptable to all voters at time t^*.

We next characterize the weighted versions of the minisum and minimax procedures in terms of the weighted distances they minimize. At the same time, we develop a simple procedure for computing all winning committees.

THEOREM 3 *A subset $p \in C$ is an MV_w committee iff it minimizes*

$$\sum_{j=1}^{\ell} w_j d(p, q^j).$$

A subset $p \in C$ is an FB_w committee iff it minimizes

$$\max_{j=1...\ell} w_j d(p, q^j).$$

Proof. For any fixed $h = 1, \ldots, k$ and any fixed $j = 1, \ldots, \ell$, define

$$\delta_h(p, q^j) = \begin{cases} w_j & \text{if } q_h^j \neq p_h \\ 0 & \text{if } q_h^j = p_h \end{cases}$$

from which it follows that

$$\sum_{j=1}^{\ell} \delta_h(p, q^j) = \begin{cases} \sum_{j=1}^{\ell} w_j q_h^j & \text{if } p_h = 0 \\ \sum_{j=1}^{\ell} w_j(1 - q_h^j) & \text{if } p_h = 1 \end{cases}$$

Therefore, using (10),

$$\sum_{j=1}^{\ell} \delta_h(p, q^j) = \begin{cases} n_w(h) & \text{if } p_h = 0 \\ 2S - n_w(h) & \text{if } p_h = 1 \end{cases}$$

But $\sum_{j=1}^{\ell} w_j d(p, q^j) = \sum_{j=1}^{\ell} \sum_{h=1}^{k} \delta_h(p, q^j) = \sum_{h=1}^{k} \sum_{j=1}^{\ell} \delta_h(p, q^j)$. It follows that $p \in C$ minimizes $\sum_{j=1}^{\ell} w_j d(p, q^j)$ iff (i) $p_h = 0$ whenever $n_w(h) < S$ and (ii) $p_h = 1$ whenever $n_w(h) > S$. This proves the first statement of the theorem.[3]

The second statement follows directly from the result of (Brams and Kilgour, 2001, Theorem 4) that the FB_w procedure determines the set of all subsets that minimize the maximum distance. ∎

---

[3] This part of the proof generalizes the proof of (Brams, Kilgour and Sanver, 2004, Proposition 4.)

We will refer to MV_w and FB_w committees as weighted minisum and weighted minimax committees, respectively. Theorem 3 and its proof provide a simple procedure for finding all such committees.

Given compressed ballot data (Q, m) and a weight, w_j, for each voted-for subset, q^j, compute a $2^k \times \ell$ matrix in which the rows represent subsets and the columns voted-for subsets. The (p, j) entry is $w_j d(p, q^j)$. Then find the sum and the maximum of the entries in each row. The rows with minimum sum correspond to the weighted-minisum committees, and the rows with the minimum maximum to the weighted-minimax committees.

We illustrate first with Example 2 using count weights. The three columns of the body of Table 3 refer to the voted-for committees (listed across the top). Below each voted-for committee is its weight—in this case, its component of the count vector. The rows of the table correspond to the possible winning committees, which in this illustration are all subsets of $C = \{1, 2, 3\}$. Each entry of the table is the weighted distance between the column subset and the row subset. The two right-most columns record the row sum and row maximum.

Table 3. Determination of winning subset(s), Example 2, with count weights.

| *Voted-for Committee:* | 100 | 110 | 101 | | |
Weight:	1	1	2	$\sum$	max
Subsets: 000	1	2	4	7	4
100	0	1	2	3*	2*
010	2	1	6	9	6
001	2	3	2	7	3
110	1	0	4	5	4
101	1	2	0	3*	2*
011	3	2	4	9	4
111	2	1	2	5	2*

The winning subsets are indicated with asterisks. The weighted-minisum committees are 100 and 101, and the weighted-minimax committees are 100, 101, and 111. These results accord exactly with those given earlier.

Note also that the table simplifies an extended procedure we recommended elsewhere (Brams, Kilgour, and Sanver, 2005): Choose either minisum and minimax as the primary criterion; use the other as a secondary criterion to

Table 4.   Determination of winning subset(s), Example 2, with proximity weights.

Voted-for Committee:	100	110	101		
Weight:	5	3	10	Σ	max
Subsets: 000	5	6	20	31	20
100	0	3	10	13	10
010	10	3	30	43	30
001	10	9	10	29	10
110	5	0	20	25	20
101	5	6	0	11*	6*
011	3	2	4	9	4
111	10	3	10	23	10

break ties. Doing so leads us to discard 111 as a minimax committee, leaving 100 and 101 as the two committees selected according to both criteria.[4]

We next analyze the same example using proximity weights. As determined by (9), these work out to be $w_1 = \frac{1}{3}$, $w_2 = \frac{1}{5}$, and $w_3 = \frac{2}{3}$. (For instance, from $q^1 = 100$ the Hamming distances to q^1, $q^2 = 110$, and $q^3 = 101$ are 0, 1, and 1, respectively; because q^1, q^2, and q^3 are selected by $m_1 = 1$, $m_2 = 1$, and $m_3 = 2$ voters, respectively, $w_1 = \frac{1}{(1\times0)+(1\times1)+(2\times1)} = \frac{1}{3}$.) To make Table 4 neater, we have cleared denominators by multiplying the weights by 15, producing $15 \times \frac{1}{3} = 5$, $15 \times \frac{1}{5} = 3$, and $15 \times \frac{2}{3} = 10$.

There is now a unique winning committee, 101, as shown by the asterisks, according to both the minisum and minimax criteria.

Of course, there may be restrictions on the size or composition of the committee to be elected. Restrictions might be imposed, for example, to fix the exact size of the committee, to establish bounds (upper, lower, or both) on its size, or to ensure that the committee contains representatives of certain subgroups of the candidate set. One important feature of our procedures is that they can be adapted to satisfy restrictions of this sort.

In fact, changing the procedure to incorporate such restrictions is easy. Simply delete from the table all rows that correspond to subsets that fail to meet the restrictions. For example, in the table above (Example 2, using proximity weights), a plausible restriction might be that the committee to be elected is to have exactly one member. If so, only the second, third, and fourth rows of the table correspond to eligible subsets; all other rows must be removed. Clearly,

---

[4]A more sophisticated approach would be to order the rows of the table using the leximin ordering—see (Fishburn, 2001) for details.

the committee selected would be 100 by the minisum criterion, and either 100 or 001 by the minimax criterion.[5]

4.   Conclusions

We have analyzed how minisum and minimax criteria can be used to elect a representative committee using approval balloting. Motivating our study is a natural link between approval balloting and committee elections: Both the ballot and the outcome of the election are subsets of the set of candidates. We have assessed minisum and minimax criteria, which give rise to procedures with several desirable properties—they are anonymous (treat all voters equally), neutral (treat all candidates equally), and symmetric (votes for and against a candidate have an equal influence on the outcome). Moreover, we have shown how both of these criteria can be adjusted using weights.

Based on the construction of tables such as Table 3 and Table 4, it is possible to draw some inferences about the complexity of the minisum and minimax criteria.[6] Compressed ballot data, (Q, m), determines the number of voters, n, the number of candidates, k, and the number of voted-for committees, ℓ. Count weights can, of course, be obtained directly from the data. To calculate all proximity weights using (9) takes approximately $\ell^3 k$ arithmetic operations.

Now suppose that the minisum or minimax procedure is to be applied using a table. The table has $\ell 2^k$ entries, each of which is a (weighted) distance determined by ℓk comparisons. Once the table is constructed, the determination of each row maximum, and the comparison of the maxima, takes relatively few steps. Thus, the complexity of either procedure, or both together, is roughly $\ell^2 k 2^k$.

Note, however, that the minisum criterion can be applied much more efficiently using (5), if it is known that no candidate received exactly $\frac{n}{2}$ votes—for example, if n is odd. Then the unique minisum committee can be determined by using (10) to calculate the weighted vote for each candidate. The required subset contains all candidates whose weighted vote exceeds the threshold S. This calculation takes on the order of ℓk arithmetic operations.

Thus, the order of complexity of the minimax procedure is always higher. In the worst case, when $\ell = n$, the minimax procedure may be exponential in the number of candidates, but it is always polynomial in the number of voters. Hence, we believe that both the minisum and minimax criteria can be the basis of practical procedures for using approval balloting in the election of a committee. Further study of their properties is merited.

---

[5] Application of the two criteria in sequence, as discussed above, would break the tie in favor of 100.
[6] Algorithms for the computation of minisum and minimax vertices on graphs are studied in (Mirchandani and Francis, 1990).

References

Brams, Steven J., and Peter C. Fishburn (1978). "Approval Voting," *American Political Science Review* 72, 3, 831–857.

Brams, Steven J., and Peter C. Fishburn (1983). *Approval Voting*, Cambridge, MA: Birkhäuser Boston.

Brams, Steven J., and Peter C. Fishburn (2005). "Going from Theory to Practice: The Mixed Success of Approval Voting," *Social Choice and Welfare* 25, 2–3, 457–474.

Brams, Steven J., and D. Marc Kilgour (2001). "Fallback Bargaining," *Group Decision and Negotiation* 10, 4, 287–316.

Brams, Steven J., D. Marc Kilgour, and M. Remzi Sanver (2004). "A Minimax Procedure for Negotiating Multilateral Treaties," in *Reasoned Choices: Essays in Honor of Hannu Nurmi*, (ed. Matti Wiberg), Turku, Finland: Finnish Political Science Association, 108–139.

Brams, Steven J., D. Marc Kilgour, and M. Remzi Sanver (2005). "A Minimax Procedure for Electing Committees," paper presented at Annual Meeting of the American Political Science Association, Washington, DC, September 1–4.

Fishburn, Peter C. (2001). "Lexicographic Orders, Utilities, and Decision Rules," *Management Science* 20, 11, 1442–1471.

Merrill, Samuel, III, and Jack H. Nagel (1987). "The Effect of Approval Balloting on Strategic Voting under Alternative Decision Rules," *American Political Science Review* 81, 2, 509–524.

Mirchandani, P.B. and R.L. Francis, editors. (1990). *Discrete Location Theory*, New York, NY: Wiley Interscience.

Rawls, John (1971). *A Theory of Justice*, Cambridge, MA: Harvard University Press.

On Some Distance Aspects in Social Choice Theory

Christian Klamler
University of Graz

Abstract   This paper investigates the relationship between Kemeny-type distance functions on the set of choice functions and the original Kemeny distances on the set of binary relations with special emphasis on the set of linear orders. First, it will be shown in what way such distances differ. Second, for the Kemeny-type distance function on the set of choice functions we will provide an explicit expression in terms of the linear orders that rationalize them.

1.   Introduction

The idea of measuring distances between binary relations has been increasingly used in areas such as social choice and computer sciences in problems of aggregation of individual rankings into a group ranking. Most prominent in that respect is the distance function devised and characterized by Kemeny (1959), which is based on the cardinality of the symmetric difference between binary relations.[1] In contrast to binary relations, choice functions are an alternative and very attractive possibility to represent (individual) preferences that are used in economic and political models. Choice functions have been intensively analysed in Aizerman and Aleskerov (1995) and Aleskerov and Monjardet (2002). Especially in aggregation problems whenever little structure is imposed on individual and/or group preferences, the use of choice functions seems compelling (Xu, 1996). In that respect, choice functions can help to avoid various paradoxes, such as Arrow's impossibility theorem (Sen, 1986). Distance functions on the set of choice functions have been used in connection to convexity issues and the aggregation of individual choice functions by Albayrak and Aleskerov (2000) and Ilyunin and Popov (1988). A characterization of such a distance function which is based on the cardinality of the symmetric differences of choice sets - and hence in the spirit of the Kemeny

---

[1] A generalization of the Kemeny distance function to a distance function on partial orders has been provided by Bogart (1973).

distance on binary relation - can be found in Klamler (2005).

The main goal of this note is to investigate the relationship between such a Kemeny-type distance function on the set of choice functions and the original Kemeny distance on the set of binary relations, where in this paper we focus on linear orders. First, we will show in what way such distances differ. Second, for the Kemeny distance on linear orders we will provide an explicit expression in terms of the rationalized choice functions, and for the Kemeny-type distance function on the set of choice functions we will provide an explicit expression in terms of the linear orders that rationalize them. It will be seen that the latter is based on the alternatives' positions in the corrsponding linear orders.

2.   Formal Framework

Let X be a finite set of m alternatives and denote the set of all non-empty subsets of X by K. A choice function on X is a function $C : K \rightarrow K$ such that for all $S \in K, C(S) \subseteq S$. The set of all choice functions is denoted by $\mathcal{C}$. A function $d : \mathcal{C} \times \mathcal{C} \rightarrow \mathbb{R}_+$ is called a distance function on set $\mathcal{C}$ if it satisfies the following three conditions:

- $d(C, C') = 0 \Leftrightarrow C = C'$

- $d(C, C') = d(C', C)$

- $d(C, C'') \leq d(C, C') + d(C', C'')$

In the following we will use the notion of "betweenness" for choice functions as stated in Albayrak and Aleskerov (2000). In particular we will say that for any $C, C', C'' \in \mathcal{C}$, choice function C' lies between C and C'', if for all $S \in K, C(S) \cap C''(S) \subseteq C'(S) \subseteq C(S) \cup C''(S)$.

Klamler (2005) characterizes a distance function on $\mathcal{C}$ using the following properties:

A1.1 $d(C, C') \geq 0$ where equality holds if and only if $C = C'$

A1.2 $d(C, C') = d(C', C)$

A1.3 $d(C, C'') \leq d(C, C') + d(C', C'')$ and equality holds if and only if C' is between C and C''.

A2 If $\tilde{C}, \tilde{C}'$ results from C, C' by a permutation of the alternatives, then $d(C, C') = d(\tilde{C}, \tilde{C}')$

A3 If two choice functions $C, C' \in \hat{\mathcal{C}}$ overlap except for a set $\bar{K} \subset K$ which is part of the domain in both choice functions, then the distance $d(C, C')$ is determined exclusively from the choice sets over $\bar{K}$.

A4 Let four choice functions $C, C', C'', C''' \in \hat{C}$ disagree only on set $T \in K$ such that for some $S \subseteq T$, $C(T) = C''(T) \cup S$, and $C'(T) = C'''(T) \cup S$. Then the distance between C and C' should be equal to the distance between C'' and C'''.

A5 The minimal positive distance is 1.

Let us now define the following distance function on C:

DEFINITION 1 *For any* $C, C' \in C$, $d_F(C, C') = \sum_{S \in K} |C(S) \Delta C'(S)|$.

Hence, d_F measures the sum of the cardinality of the symmetric differences of all choice sets in the two choice functions. The following result has been proved in Klamler (2005).[2]

THEOREM 2 *A distance function d on C is equal to d_F if and only if it satisfies the axioms A1-A5.*

Another, more widely used way of representing preferences, is to take binary relations $R \subset X \times X$. We will use xRy for saying that x is at least as good as y (or $x \geq y$). Using similar axioms as those in the above characterization, Kemeny (1959) characterized a distance function on the set of binary relations. In the following we will restrict ourselves to the set of all linear orders, $\mathcal{L}$, i.e. all complete, antisymmetric and transitive binary relations on X.

DEFINITION 3 *A function $d_K : \mathcal{L} \times \mathcal{L} \to R_+$ is the Kemeny distance if for all $R, R' \in \mathcal{L}$, $d_K(R, R') = |R \Delta R'|$.*

3. Choice Functions Versus Binary Relations

In this section we want to investigate whether the distance between choice functions, measured by d_F, corresponds to the distance between the preferences that rationalize them, measured by d_K.[3] We say that a choice function C is *linearly rationalizable* if there exists an $R \in \mathcal{L}$ such that for all $S \in K$, $C(S) = max_R S$, where $max_R S = \{x \in X : \forall y \in X, xRy\}$. It should be clear that this leads to the following:

REMARK 4 **a** *If $C \in C$ is linearly rationalized by $R \in \mathcal{L}$, then $xRy \Leftrightarrow C(\{x, y\}) = \{x\}$.*

b *If $C, C' \in C$ are linearly rationalized by $R, R' \in \mathcal{L}$, respectively, then $C \equiv C' \Leftrightarrow R \equiv R'$.*

---

[2] Actually the proof is based on the larger set of "quasi" choice functions, i.e. choice functions where empty choice sets are possible.
[3] On rationalization of choice functions see Suzumura (1983).

From Remark 4(b) we see that there is a one-to-one correspondence, ϕ, between the set of linear orders and the set of choice functions rationalized by them.

EXAMPLE 5 *Let $R_1, R_2, R_3 \in \mathcal{L}$ be as stated in Table 1 (where less preferred alternatives are in lower rows).*

Table 1.   Linear orders with 3 alternatives.

R_1	R_2	R_3
x	y	x
y	x	z
z	z	y

The corresponding choice functions C_1, C_2, C_3 are given in Table 2 (for notational convenience the value of C on singletons will be omitted throughout the paper).

Table 2.   Choice functions rationalized by the linear orders in Table 1.

	$C_1(.)$	$C_2(.)$	$C_3(.)$
X	x	y	x
xy	x	y	x
xz	x	x	x
yz	y	y	z

Using our previously defined distance functions d_K and d_F, the distances between the above preferences are as follows:

$$d_K(R_1, R_2) = 2; d_K(R_1, R_3) = 2$$
$$d_F(C_1, C_2) = 4; d_F(C_1, C_3) = 2$$

PROPOSITION 6 *There exist binary relations $R_1, R_2, R_3 \in \mathcal{L}$ and corresponding choice functions $C_1, C_2, C_3 \in \mathcal{C}$, such that $d_K(R_1, R_2) \geq d_K(R_1, R_3)$ and $d_F(C_1, C_2) < d_F(C_1, C_3)$.*

Proof. See example 1. ∎

It seems quite remarkable that such similar concepts of measuring distance provide different results. Actually for four or more alternatives the distances could switch completely as stated in the following proposition:

PROPOSITION 7 *If $m \geq 4$, there exist binary relations $R_1, R_2, R_3 \in \mathcal{L}$ and corresponding choice functions $C_1, C_2, C_3 \in \mathcal{C}$, such that $d_K(R_1, R_2) > d_K(R_1, R_3)$ and $d_F(C_1, C_2) < d_F(C_1, C_3)$.*

Proof.

Let us start with the case $m = 4$ and let there be $R_1, R_2, R_3 \in \mathcal{L}$ given as in Table 3.

Table 3. Linear orders with 4 alternatives.

R_1	R_2	R_3
x	y	x
y	x	w
z	z	y
w	w	z

The corresponding choice functions that are rationalized by those linear orders are stated in Table 4.

Table 4. Choice functions rationalized by linear orders in Table 3

X	$C_1(.)$	$C_2(.)$	$C_3(.)$
xyz	x	y	x
xyw	x	y	x
xzw	x	x	x
yzw	y	y	w
xy	x	y	x
xz	x	x	x
xw	x	x	x
yz	y	y	y
yw	y	y	w
zw	z	z	w

In this example the distances between the binary relations are $d_K(R_1, R_2) = 2$ and $d_K(R_1, R_3) = 4$, but the distances between their corresponding choice functions are $d_F(C_1, C_2) = 8$ and $d_F(C_1, C_3) = 6$. Hence, what is considered of larger distance in space $\mathcal{L}$ is considered of lower distance in space $\mathcal{C}$.

For any situation with $m > 4$ we can now introduce additional alternatives such that the switches in the binary relations are maintained as in the above example. Assume set $A \subset X$ being a linearly ordered set of alternatives in X and consider $a \in X$. A general binary relation would be as in Table 5, where alternatives above A are strictly preferred to all alternatives in A and alternatives below A are strictly preferred by all alternatives in A.

Hence the distance between R_1 and R_2 is $d_K(R_1, R_2) = 2$ whereas $d_K(R_1, R_3) = 4$. However, as R_1 and R_2 only differ in the first two alternatives, this means that in the respective choice functions there is a change in the choice set

Table 5.   Linear orders with m alternatives.

R_1	R_2	R_3
x	y	x
y	x	y
A	A	A
a	a	w
z	z	a
w	w	z

from x to y in 2^{m-2} sets leading to a distance $d_F(C_1, C_2) \geq 16$ for $m \geq 5$. On the other hand the change of alternative w from bottom position to position $m - 2$ leads to a change in 3 choice sets and hence to a distance $d_F(C_1, C_3) = 6$. ∎

What could be the reason for this observation? It turns out that in using distance function d_F, switches between alternatives which are higher up in the corresponding ranking get more weight than switches at the lower end of the ranking. Switching the two bottom ranked alternatives only changes the choice of one set (namely that of the two alternatives) whereas switching the two top ranked alternatives changes the choice of 2^{m-2} sets (namely all those in which both alternatives are contained). This might indeed be a reasonable way of thinking about distances especially as in many voting situations alternatives lower down the ranking are not considered at all. The different weights assigned to switches between alternatives depending on their positions in the ranking will be of importance in the analysis of the distance functions.

4.   Distance Functions

The previous section has illuminated the differences between d_K and d_F. Now, the question arises whether we can explicitly express $d_K(R, R')$ in terms of the corresponding rationalized choice functions $\phi(R)$ and $\phi(R')$ and whether we can explicitly express $d_F(C, C')$ in terms of the rationalizing binary relations $\phi^{-1}(C)$ and $\phi^{-1}(C')$.

From Remark 4(b) we know that if $C \in \mathcal{C}$ is linearly rationalized by $R \in \mathcal{L}$ then $xRy \Leftrightarrow C(\{x, y\}) = \{x\}$. Hence it is obvious that for all $R, R' \in \mathcal{L}$ such that xRy and $yR'x$, the Kemeny distance between R and R' restricted to set $\{x, y\}$ is 2 and this is equal to the cardinality of the symmetric difference of the corresponding choices from $\{x, y\}$, i.e. $|C(\{x, y\}) \Delta C'(\{x, y\})| = |\{x\} \Delta \{y\}| = 2$. Hence this leads to the obvious corollary:

COROLLARY 8 *For any $R, R' \in \mathcal{L}$ with corresponding $C, C' \in \mathcal{C}$,*

$$d_K(R, R') = |R \Delta R'| = \sum_{S \in K_2} |C(S) \Delta C'(S)|$$

where $K_2 = \{T : |T| = 2\}$.

It is by far less obvious to explicitly express $d_F(C, C')$ in terms of $\phi^{-1}(C)$ and $\phi^{-1}(C')$. As discussed in the previous section, d_F takes into account the positions of the alternatives in the corresponding rankings. Hence it seems clear that in addition to the number of switches in pairs of alternatives also positional information about those alternatives will be required. Therefore we will introduce the following notion of switching:[4]

Let $R, R' \in \mathcal{L}$. Since R and R' are both complete and antisymmetric it follows that $(x, y) \in R \backslash R' \Leftrightarrow (y, x) \in R' \backslash R$. We call a pair $(x, y) \in R$ an $R - R'$-transposition if $(y, x) \in R'$.[5] Moreover, $R - R'$-transpositions (x, y) can be grouped by x. Let $\tau(x) \equiv \tau_{R,R'}(x) = |\{y \in X | (x, y) \in R$ and $(y, x) \in R'\}|$ be the number of transpositions involving x (where $\tau(x)$ might be zero). Let $p_R : X \to \{1, 2, ..., m\}$ be the ranking of X w.r.t. R, i.e.,

$$(x, y) \in R \text{ and } x \neq y \Rightarrow p_R(x) < p_R(y).$$

This enables us to make the following claim:

CLAIM 9 *Let $R, R' \in \mathcal{L}$ with corresponding choice functions $C, C' \in \mathcal{C}$ and, for $x \in X$, let $K_x \equiv \{S \in K | C(S) = \{x\}\}$. Then*

$$\sum_{S \in K_x} |C(S) \Delta C'(S)| = 2 \left(2^{m - p_R(x)} - 2^{m - p_R(x) - \tau(x)} \right)$$

Proof. Obviously, for C rationalized by R, $|K_x| = 2^{m - p_R(x)} - 1$, namely the number of sets in which x is chosen. By the same argument, the number of sets in which x is chosen w.r.t. R' is $2^{m - p_{R'}(x)} - 1$. As we restrict ourselves to alternatives for which $p_R(x) > p_{R'}(x)$ it follows from $p_{R'} = p_R - \tau(x)$ that a change in the choice sets occured in exactly $2^{m - p_R(x)} - 2^{m - p_R(x) - \tau(x)}$ sets. As $R, R' \in \mathcal{L}$, and hence $|C(S) \Delta C'(S)|$ is either zero or two, the above claim follows. ∎

Hence a straightforward corollary is the following:

---

[4] I am very grateful to a referee for suggesting this idea and herewith making the results much clearer and simpler.

[5] Notice that the Kemeny distance $d_k(R, R')$ between R and R' is twice the number of $R - R'$-transpositions.

COROLLARY 10

$$d_F(C, C') = \sum_{x \in X} \sum_{S \in K_x} |C(S) \Delta C'(S)| = 2 \sum_{i=1}^{m-1} \left(2^{m-i} - 2^{m-i-\tau(x_i)} \right)$$

Obviously, this expresses $d_F(C, C')$ as a function of $\tau \equiv \tau_{R,R'}$ and hence of R and R'.

5.   Summary

This paper has investigated the difference between Kemeny-type distance functions based on the set of all choice functions and the original Kemeny distance on binary relations. We have illuminated the differences between those distances. Moreover, both distance concepts seem reasonable as they differ only by the weight attached to the alternatives according to their positions in the preference rankings. Finally, we have provided explicit statements of those distance functions using their counterparts, namely linear orders and choice functions, respectively.

Acknowledgements

I am grateful to Fuad Aleskerov, Hannu Nurmi, Daniel Eckert and an anonymous referee for their helpful comments. Especially the anonymous referee has provided invaluable input in simplifying and clarifying definitions and proofs compared to a previous draft. All remaining errors are of course mine.

References

Aizerman, M., Aleskerov, F., 1995. Theory of Choice, Elsevier, Amsterdam.

Albayrak, S.R., Aleskerov, F., 2000. Convexity of Choice Function Sets. Bogazici University Research Paper, ISS / EC-2000-01, 2000.

Aleskerov, F., Monjardet, B., 2002. Utility Maximization, Choice and Preference, Springer, Berlin.

Bogart, K.P., 1973. Preference Structures I: Distances between transitive perference relations. Journal of Mathematical Sociology, 3, 49-67.

Ilyunin, O.K., Popov, B.V. and L.N. El'kin, 1988. Majority Functional Operators in Voting Theory. Automation and Remote Control. 7, 137-145.

Kemeny, J., 1959. Mathematics without numbers. Daedalus 88, 571-591.

Klamler, C., 2005. Choice Functions, Binary Relations and Distances, mimeo, Graz University.

Sen, A., 1986. Social choice theory. In: Arrow., K.J., Intrilligator, M.D. (Eds.), Handbook of Mathematical Economics, vol. III, North Holland, Amsterdam, pp. 1073-1181, Chapter 22.

Suzumura, K., 1983. Rational choice, collective decisions, and social welfare. Cambridge University Press, Cambridge.

Xu, Y., 1996. Non Binary Social Choice: A Brief Introduction. in: Schofield, N. (ed.), Collective Decision-Making: Social Choice and Political Economy, Kluwer, Boston.

Algorithms for Biproportional Apportionment

Sebastian Maier
Institut für Mathematik, Universität Augsburg

Abstract For the biproportional apportionment problem two algorithms are discussed, that are implemented in the Augsburg BAZI program, the alternating scaling algorithm and the tie-and-transfer algorithm of Balinski and Demange (1989b). The goal is to determine an integer-valued apportionment matrix that is "proportional to" a matrix of input weights (e.g. vote counts) and that at the same time achieves prespecified row and column marginals. The alternating scaling algorithm finds the solution of most of the practical problems very efficiently. However, it is possible to create examples for which the procedure fails. The tie-and-transfer algorithm converges always, though convergence may be slow. In order to make use of the benefits of both algorithms, a hybrid version is proposed.

Keywords: Biproportional divisor method; biproportional rounding algorithm; discrete alternating scaling; tie-and-transfer algorithm; BAZI.

1. Introduction

The Zurich Canton parliament is composed of seats that represent electoral districts as well as political parties (Pukelsheim, 2004b; Balinski and Pukelsheim, 2006; Pukelsheim, 2006). Each district $i = 1, \ldots, k$ is represented by a number of seats r_i proportional to its population, and each political party $j = 1 \ldots l$ gets c_j seats proportional to its total number of votes. The vote count in district i of party j is denoted by v_{ij}. Altogether the vote counts are assembled into a vote matrix $V \in \mathbb{N}^{k \times l}$ (see Box on page 106).

Box: Biproportional divisor method with standard rounding (Neues Zürcher Zuteilungsverfahren).

The seats per district are apportioned in the middle of the legislature period on the basis of the population counts. Party seats are allocated on election day on the basis of the total party ballots in the whole electoral region. As the Zurich electoral law provides each voter with as many ballots as the district has seats, we need to compute the support for a party in a district. This is done by dividing the raw data that are returned by the polling stations by the district magnitude, and by rounding the resulting quotient to the closest integer. For each party, these district support sizes are summed over all districts leading to the overall support size for a party. The support size may be interpreted as number of people supporting a party. It is used to compute the superapportionment, this is, the allocation of the seats to the parties across the whole electoral region.

The final step is the subapportionment, the allocation of the seats to the parties within the districts. It provides a two-way proportionality, achieving the prespecified district magnitudes and the party seats. To compute the apportionment we need two sets of divisors, the district divisors and the party divisor. Each vote count of a party in a district is divided by its corresponding district divisor and party divisor; this quotient is rounded in the usual way to obtain the seat-number. A more detailed description of the Zurich apportionment method can be found in Balinski and Pukelsheim (2006) and Pukelsheim (2006).

Zurich City Parliament election of 12 February 2006, Superapportionment:

	SP	SVP	FDP	Greens	CVP	EVP	AL	SD	City divisor
Support size	23180	12633	10300	7501	5418	3088	2517	1692	530
Seats 125	44	24	19	14	10	6	5	3	

Zurich City Parliament election of 12 February 2006, Subapportionment:

	SP	SVP	FDP	Greens	CVP	EVP	AL	SD	District-
	44	24	19	14	10	6	5	3	divisor
125									
"1+2" 12	28518-4	15305-2	21833-3	12401-2	7318-1	2829-0	2413-0	1651-0	7000
"3" 16	45541-7	22060-3	10450-1	17319-3	8661-1	2816-0	7418-1	3173-0	6900
"4+5" 13	26673-5	8174-2	4536-1	10221-2	4099-1	1029-0	9086-2	1406-0	5000
"6" 10	24092-4	9676-1	10919-2	8420-1	4399-1	3422-1	2304-0	1106-0	6600
"7+8" 17	61738-5	27906-2	51252-5	25486-2	14223-1	10508-1	5483-1	2454-0	11200
"9" 16	42044-6	31559-4	12060-2	9154-1	11333-1	9841-1	2465-0	5333-1	7580
"10" 12	35259-4	19557-3	15267-2	9689-1	8347-1	4690-1	2539-0	1490-0	7800
"11" 19	56547-6	40144-4	19744-2	12559-1	14762-2	11998-2	3623-1	6226-1	9000
"12" 10	13215-3	10248-3	3066-1	2187-1	4941-1	0-0	429-0	2078-1	4000
Partydivisor	1.006	1.002	1.01	0.97	11333-1	0.88	0.8	1	

Table entries are of the form v-a, where v is the number of party votes in the district and a is the number of seats apportioned to that party's list in the district. The party ballot v is divided by the associated district and party divisors, and then rounded in the standard way to obtain a. In district "1+2" the Greens had 12401 ballots and were awarded by 2 seats, since $12401/(7000 \times 0.97) = 1.83 \nearrow 2$.

This leads to the following proportional matrix problem (cf. Balinski and Demange 1989a,b; Pukelsheim, 2004; Balinski and Pukelsheim, 2006; Pukelsheim, 2006). Find a matrix apportionment $A \in \mathbb{N}^{k \times l}$ and row divisors ρ_i, $i = 1 \ldots k$ and column divisors γ_j, $j = 1 \ldots l$ which satisfy the following conditions:

$$a_{ij} = \left[\frac{v_{ij}}{\rho_i \cdot \gamma_j} \right]_s \tag{1}$$

$$a_{i+} = \sum_{j \leq l} a_{ij} = r_i, \ i = 1, \ldots, k \tag{2}$$

$$a_{+j} = \sum_{i \leq k} a_{ij} = c_j, \ j = 1, \ldots, l. \tag{3}$$

The rounding $[x]_s$ of a positive number $x \in [n, n+1]$, $n \in \mathbb{N}$ depends on a dividing point $s(n) \in [n, n+1]$. We have $[x]_s = n+1$, if $x > s(n)$, and $[x]_s = n$, if $x < s(n)$. In the case x hits the signpost $s(n)$, x can be either rounded down or up. Thus the commonly known divisor methods for vector apportionments are extended to the matrix case. Matrix apportionments which are computed using divisor methods share nice properties, e.g. uniformity and homogeneity, and are unique up to ties (Balinski and Demange 1989a,b).

Since the matrix problem cannot be solved in one step, we need iterative procedures. In Section 2 we review two algorithms for computing the matrix apportionment. In Section 3 we investigate the runtime and the error functional. This leads us to suggest merging the advantages of both algorithms to a hybrid algorithm in Section 4.

2. Algorithms

The two algorithms to be reviewed are the alternating scaling algorithm, a discrete version of the commonly known iterative proportional fitting algorithms from statistics, and the tie-and-transfer algorithm proposed in (Balinski and Demange, 1989b).

For both algorithms, a measure for the improvement is the error count in step t which corresponds to an interim apportionment $A(t)$:

$$f(t) := \frac{1}{2} \sum_{i \leq k} |a_{i+}(t) - r_i| + \frac{1}{2} \sum_{j \leq l} |a_{+j}(t) - c_j|$$

This is the number of seat transfers necessary to achieve the solution, that is, the number of "wrong" allocations within the apportionment matrix. The procedure stops as soon as the error count is zero.

Before starting the computation, existence of a solution can be checked by using a max-flow min-cut algorithm (Joas, 2005). For the continuous case, existence is investigated for example in Bacharach (1970) or Pretzel (1980). In the sequel, we assume existence. Also we do not pay attention to multiplicities that are possible when the scaled weights hit a signpost.

2.1 Alternating Scaling Algorithm (AS)

The continuous alternating scaling algorithm was proposed by Deming and Stephan (1940) and has many applications in statistics, such as fitting contingency tables or fitting loglinear models (Fienberg and Meyer, 1983). This procedure, also known as the RAS algorithm, is extensively studied in literature (Ireland and Kullback, 1968; Marshall and Olkin, 1968; Bacharach, 1970). A more extensive and detailed overview on the literature can be found in Balinski and Pukelsheim (2006). The discrete version of the alternating scaling procedure and its properties is described in Pukelsheim (2004). The idea is to solve the vector problem for rows in odd steps and for columns in even steps. Hence, either rows or columns are fitted. If the rows are fitted, errors may be left in the columns, and if the columns are fitted, errors may be left in the rows. Thus the matrix problem is reduced to many vector problems, either by solving a set of row problems, or by solving a set of column problems. To solve such a vector problem, the algorithm described in Dorfleitner and Klein (1999) is used in updating the divisors in each step. The algorithm succeeds in presenting row and column divisors fulfilling (1) – (3). A Java implementation of the following algorithm can be downloaded from www.uni-augsburg.de/bazi.

Algorithm (Discrete Alternating Scaling)

(0) *Initialize start divisors $P_i(0) = 1, i = 1 \ldots, k$ and $\Gamma_j(1) = 1, j = 1, \ldots, l$. At every step t, the scaled weights will be of the form $v_{ij}(t) = \frac{v_{ij}}{P_i(t) \cdot \Gamma_j(t)}$.*

(i) *For odd steps, find row divisors $\rho_i(t)$ such that, with updated divisors $P_i(t) = \rho_i(1)\rho_i(3) \ldots \rho_i(t)$, the apportionment $a_{ij}(t) = [v_{ij}(t)]_s$ satisfies (2).*

(ii) *For even steps, find column divisors $\gamma_j(t)$ such that, with updated divisors $\Gamma_j(t) = \gamma_j(2)\gamma_j(4) \ldots \gamma_j(t)$, the apportionment $a_{ij}(t) = [v_{ij}(t)]_s$ satisfies (3).*

The algorithm terminates successfully after finitely many steps, T say, when (1) – (3) are satisfied with divisors $\rho_i = P_i(T)$ and $\gamma_j = \Gamma_j(T)$.

Note that row divisors are updated only in odd steps, and column divisors in even steps. Therefore the updated divisors are of the form

$$P_i(t) = \rho_i(1)\rho_i(3)\cdots\rho_i\left(2\lceil\frac{t}{2}\rceil - 1\right), \ \Gamma_j(t) = \gamma_j(2)\gamma_j(4)\cdots\gamma_j\left(2\lfloor\frac{t}{2}\rfloor\right).$$

2.2   Tie-and-Transfer (TT)

The tie-and-transfer algorithm was first described in Balinski and Demange (1989b), in a more general form dealing with inequality constraints for rows and columns. The case of equality constraints can be found in Balinski and Rachev (1997). The main idea is to transform the biproportional problem into a bipartite graph. This graph is used to find a feasible flow corresponding to a biproportional apportionment (Balinski and Demange, 1989b; Balinski and Rachev, 1997; Zachariasen, 2006).

Algorithm (Tie-and-Transfer)

(0) Start with an initial apportionment exhausting the housesize. This initial apportionment is obtained by fitting all columns as proposed in Balinski and Rachev (1997). Then the following labelling procedure is established.

(i) Either determine subsets of rows and columns to modify the row and column divisors. The modification is done in such a way that at least one more of the rescaled weights is either scaled down to the previous signpost, or scaled up to the next signpost. This means that the rescaled weight can be rounded either down or up without affecting the divisors.

(ii) Else determine a path from an underrepresented row to an overrepresented row in the graph. The path is along an interim apportionment on rescaled weights hitting a signpost alternating being rounded down and rounded up. Along this path the direction of rounding is changed, that is, rescaled weights which were rounded down will now be rounded up, and rescaled weights which were rounded up will now be rounded down. This procedure increases the number of seats in the underrepresented row and decreases the number of seats in the overrepresented row. The net effect of the transfer is a decrease of the error function by exactly one unit. The transfer does not affect any other row or column sums.

The algorithm will terminate after finitely many steps.

Note that the apportionment is only modified on arcs that correspond to rescaled weights on signposts. This ensures at every step an apportionment which can be obtained by the current divisors. The error count decrease is

Table 1. Runtimes, iterations and starting errors: AS seems to perform better than TT, except for examples MOC3T and MOD3T.

	Tie-and-Transfer		Alternating Scaling	
	time	start error	time	iterations
	(sec.)	(count)	(sec.)	(count)
KRW1995	2	30	1	11
KRW1999	2	29	2	75
KRW2003	2	29	1	18
MOB3T	6	499	2	177
MOC3T	6	500	25	2294
MOD3T	6	500	34	3472
MOB3M	7674	499092	5	395
MOC3M	7756	499902	92	6535
MOD3M	7043	499999	112	9468

$o(lkf(0))$ (Balinski and Demange, 1989b). In contrary to the alternating scaling algorithm, the tie-and-transfer-algorithm in every step either modifies the divisors, or changes the apportionment.

3.   Properties and Data

The BAZI-program (www.-augsburg.de/bazi) includes not only various apportionment methods, but also an extensive data-base. This data-base includes examples for vector problems and for matrix problems, empirical examples, and academic ones. To study the performance of the algorithms on empirical data, we choose the last three elections for the Zurich Canton parliament, KRW1995, KRW1999, KRW2003. There are 18 districts, and 13 or 14 participating parties. The housesize is 180, the district magnitudes vary from 4 to 16, the parties get 1 to 55 seats.

We also study 3×3 weight matrices that are motivated by the literature on the continuous iterative proportional fitting algorithm (Marshall and Olkin, 1968). These matrices have large "housesizes" of 3000 (MOB3T, MOC3T, MOD3T), or 3000000 (MOB3M, MOC3M, MOD3M).

Table 1 summarizes the observed runtimes of the computation. Since the error count function is linearly decreasing for the tie-and-transfer algorithm, only the starting error count is given. For the alternating scaling algorithm the number of iterations is shown. Both algorithms are quite fast for the Zurich Canton Parliament Election data, taking about 1 to 2 seconds. For MOB3T both algorithms are very fast, but for MOC3T and MOD3T alternating scaling takes about four to six times longer than tie-and-transfer. The three examples

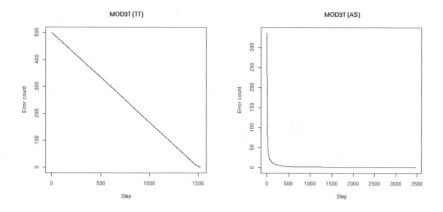

Fig. 1.   Error count for MOD3T: For TT, the decrease is constant. For AS, the decrease is fast at the beginning and slow towards the end. It reaches 0 after 3472 iterations.

with housesizes 3000000 exhibit an extreme behavior: tie-and-transfer takes hours, while the alternating scaling algorithm ends within seconds or minutes. To explain this behavior, we take a closer look at the development of the error counts.

Figure 1 shows the error count function for MOD3T for the tie-and-transfer algorithm on the left hand side, and for the alternating scaling algorithm on the right hand side. The running times are 6.5 seconds for tie-and-transfer, and 34.4 seconds for alternating scaling. For the alternating scaling algorithm, the decrease is very fast in the beginning, but very slow towards the end. It takes 225 steps for a decrease from three to two, 511 steps from two to one, and another 2200 steps to end at zero. It is known from Fienberg and Meyer (1983) that the convergence of the continuous iterative scaling procedure may be very slow. The tie-and-transfer algorithm shows the predicted linear behavior, by reducing the error count one by one. Starting with an error of 500, it is faster than alternating scaling which looses a lot of time towards the end.

Figure 2 conveys the same information. A constant decrease for the tie-and-transfer algorithm, but it has to start with an error count of 499092, and it takes rather long (about 2 hours) to work this down to zero. Alternating scaling is fast, but again the last steps take excessively long. It takes 491 steps for the decrease from three to two, 46 steps from two to one, and another 1577 steps from one to zero.

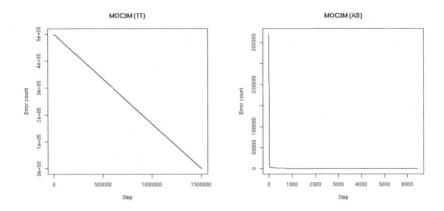

Fig. 2.   Error count for MOC3M: The decrease is constant for TT with starting error 499902. For AS, the decrease is fast at the beginning and slow towards the end, reaching 0 after 653 iterations.

4.   Hybrid Algorithm

On the basis of these examples we may summarize the runtime properties of the two algorithms: the alternating scaling algorithm is very fast in the beginning, but may take very long towards the end. The tie-and-transfer algorithm processes the error count one by one, even if there is a big error count left.

To combine the advantages of both algorithms, we suggest a hybrid version, starting with the alternating scaling algorithm and finishing with the tie-and-transfer algorithm. The challenge is to implement an appropriate time to switch.

At the end of the column adjustment there is a check for a switch of the method. We have experimented with two switching rules.

1 *Fast switch:* The error count stays the same for two iterations.

2 *Adaptive switch:* The error count stays the same as many iterations as it has digits, e.g. six iterations for an error count of 499902.

Figure 3 shows the error count decrease of the hybrid algorithm with adaptive switch for MOD3T and MOC3M. The decrease is fast in the beginning on the left hand side of the vertical line that marks the switching point. After the switch on the right hand side the error count function is linear like the error count function of the tie-and-transfer algorithm. To compare the decrease of the hybrid algorithm, the decrease of the alternating scaling algorithm is also plotted. The error count function of the hybrid algorithm decreases faster than that for the alternating scaling algorithm.

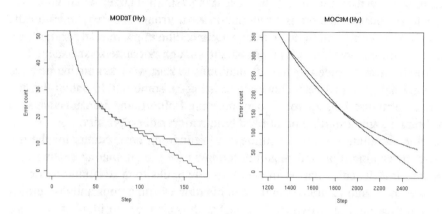

Fig. 3. Error count for MOD3T and MOC3M using the hybrid algorithm with adaptive switch: On the left hand side of the vertical line the decrease of AS can be seen. After the switching point the error count decreases linearly using the hybrid algorithm.

Table 2. Runtimes, iterations and error counts for the the tie-and-transfer algorithm, the alternating scaling algorithm, and the hybrid algorithm: The hybrid algorithm performs always better than the other proposed algorithms.

	Tie-and-Transfer		Alternating Scaling		Fast Sw. Hybrid		Adapt. Sw. Hybrid	
	time	start error	time	iter.	time	iter.+err. c.	time	iter.+err. c.
	(sec.)	(count)	(sec.)	(count)	(sec.)	(count)	(sec.)	(count)
KRW1995	2	30	1	11	1	6 + 2	1	6 + 2
KRW1999	2	29	2	75	2	6 + 1	2	6 + 1
KRW2003	2	29	1	18	2	6 + 2	2	6 + 2
MOB3T	6	499	2	177	1	52 + 20	1	52 + 20
MOC3T	6	500	25	2294	0	24 + 4	0	24 + 4
MOD3T	6	500	34	3472	1	52 + 25	1	52 + 25
MOB3M	7674	499092	5	395	4	270 + 24	4	282 + 17
MOC3M	7756	499902	92	6535	24	832 + 709	26	1380 + 317
MOD3M	7043	499999	112	9468	51	922 + 1806	40	1934 + 571

Table 2 summarizes the runtime improvements of the hybrid algorithm. In each line there is a runtime decrease. For empirical examples both switching rules have the same effect, because of the small number of error counts. For larger housesizes there is a runtime decrease from about two hours for the tie-and-transfer algorithm down to one minute. The improvement for the alternating scaling algorithm is also substantial, though not quite so spectacular.

Another way to speed up the calculation is to start with continuous iterative proportional fitting (IPF) and switch according to some rule to one of the discrete algorithms. This approach is proposed in Balinski and Demange (1989b) for finding a good starting point for the tie-and-transfer algorithm.

Table 3 summarizes running times and remaining error counts for the tie-and-transfer algorithm and required iterations for the alternating scaling procedure for a switching barrier of one. Table 4 applies for a switching barrier of ten. For the Canton Zurich data the combination of continuous and discrete alternating scaling is approximately as fast as the discrete algorithm. If there is a higher error count left, e.g. for MOC3T, MOC3M, MOD3T, and MOD3M with barrier ten, the disadvantage of slow convergence towards the end becomes visible. Using barrier one, the combination of continuous iterative proportional fitting and alternating scaling is quite fast. Using iterative proportional fitting together with tie-and-transfer together is a little slower than the other combination (IPF – AS) for the empirical examples. For the artificial examples, it has a very good performance. After the initial fitting of columns, there is no or only a small error count left for barrier one. Especially for MOC3M and MOD3M there are 369 and 445 alternating scaling steps necessary to reduce the error count to zero, but this is processed very fast by the tie and transfer algorithm.

In conclusion we find that the hybrid algorithm performs better than one of the proposed methods alone The choice of the switching rule has practically no influence, hence we decided to implement the fast switching rule in BAZI. The use of the continuous iterative proportional fitting procedure at the beginning would be another option. Another proposol to improve upon the tie-and-transfer algorithm, and a detailed investigation of the runtime properties of the algorithms can be found in Zachariasen (2006).

Acknowledgments

The author is grateful for the support of the Deutsche Forschungsgemeinschaft. I thank Michel Balinski, Bruno Simeone and an anonymous referee for all their helpful comments. A special thank to Friedrich Pukelsheim for his valuable advice and continuous support.

Table 3. Runtimes, iterations and error counts for continuous hybrid algorithm using a switching barrier of one: Starting with IPF speeds up the computation. For larger housesizes switching to TT performs better.

	TT time	AS time	IPF–TT (1) time	err. c.	IPF–AS (1) time	iter.
	(sec.)	(sec.)	(sec.)	(count)	(sec.)	(count)
KRW1995	2	1	1.3	6	0.9	16
KRW1999	2	2	2.1	8	1.3	39
KRW2003	2	1	1.6	6	0.7	11
MOB3T	6	2	0.2	1	0.2	9
MOC3T	6	25	0.8	0	0.8	2
MOD3T	6	34	1.4	0	1.3	1
MOB3M	7674	5	0.2	0	0.2	2
MOC3M	7756	92	2.1	1	5.8	369
MOD3M	7043	112	3.2	1	7.5	445

Table 4. Runtimes, iterations and error counts for continuous hybrid algorithm using a switching barrier of ten: Starting with IPF speeds up the computation. Switching to AS the disadvantage of slow convergence towards the end becomes visible again.

	TT time	AS time	IPF–TT (10) time	err. c.	IPF–AS (10) time	iter.
	(sec.)	(sec.)	(sec.)	(count)	(sec.)	(count)
KRW1995	2	1	1.3	6	0.7	16
KRW1999	2	2	2	8	1.15	32
KRW2003	2	1	1.8	6	0.76	9
MOB3T	6	2	0.2	5	1	82
MOC3T	6	25	0.2	5	24.76	2286
MOD3T	6	34	0.4	5	30.26	3105
MOB3M	7674	5	0.3	5	0.92	74
MOC3M	7756	92	1.6	5	27.95	2257
MOD3M	7043	112	2.4	5	30.8	2806

References

Bacharach, Michael (1970): *Biproportional Matrices & Input-Output Change*. Cambridge UK.

Balinski, Michel L. and Demange, Gabrielle (1989a): An axiomatic approach to proportionality between matrices. *Mathematics of Operations Research*, 14:700–719.

Balinski, Michel L. and Demange, Gabrielle (1989b): Algorithms for proportional matrices in reals and integers. *Mathematical Programming*, 45:193–210.

Balinski, Michel L. and Pukelsheim, Friedrich (2006): Matrices and politics. In E. Liski, S. Puntanen J. Isotalo, and G. P. H. Styan, editors, *Festschrift for Tarmo Pukkila on His 60th Birthday*. Department of Mathematics, Statistics, and Philosophy: Tampere.

Balinski, Michel L. and Rachev, Svetlozar T. (1997). Rouding proportions: Methods of rounding. *Mathematical Scientist*, 22:1–26.

Deming, W. Edwards and Stephan, Frederick F. (1940). On a least squares adjustment of a sample frequency table when the expected marginal totals are known. *Annals of Mathematical Statistics*, 11:427–440.

Dorfleitner, Gregor and Klein, Thomas (1999). Rounding with multiplier methods: An efficient algorithm and applications in statistics. *Statistical Papers*, 20:143–157.

Fienberg, Stephan S. and Meyer, Michael M. (1983): Iterative proportional fitting. In *Encyclopedia of Statistical Sciences*, volume 4, pages 275–279. John Wiley & Sons.

Ireland, C. T. and Kullback, S. (1968): Contingency tables with given marginals. *Biometrika*, 55(1):179–188.

Joas, Bianca (2005): A graph theoretic solvability check for biproportional multiplier methods. Diploma Thesis, Institute of Mathematics, University of Augsburg.

Marshall, Albert W. and Olkin, Ingram (1968) Scaling of matrices to achieve specified row and column sums. *Numerische Mathematik*, 12:83–90.

Pretzel, Oliver (1980): Convergence of the interative scaling procedure for non-negative matrices. *Journal of the London Mathematical Society*, 21:379–384.

Pukelsheim, Friedrich (2004): BAZI - a java programm for proportional representation. In *Oberwolfach Reports*, volume 1, pages 735–737.

Pukelsheim, Friedrich (2006): Current issues of apportionment methods. In Bruno Simeone and Friedrich Pukelsheim, editors, *Mathematics and Democracy. Recent Advances in Voting Systems and Collective Choice*. New York, 2006.

Pukelsheim, Friedrich and Schuhmacher, Christian (2004): Das neue Zürcher Zuteilungsverfahren bei Parlamentswahlen. *Aktuelle Juristische Praxis - Pratique Juridique Actuelle*, 13: 505–522.

Zachariasen, Martin (2006): Alorithmic aspects of divisor-based biproportional rounding. Typescript, April 2006.

Distance from Consensus:
A Theme and Variations [*]

Tommi Meskanen[1], Hannu Nurmi[2]

[1]Department of Mathematics, University of Turku

[2]Department of Political Science, University of Turku

Abstract   Social choice theory deals with aggregating individual opinions into social choices. Over the past decades a large number of choice methods have been evaluated in terms of various criteria of performance. We focus on methods that can be viewed as distance minimizing ones in the sense that they can be analyzed in terms of a goal state of consensus and the methods themselves can be seen as minimizing the distance of the observed profile from that consensus. The methods, thus, provide a way of measuring the degree of disagreement prevailing in the profile.

Keywords: Voting systems, metrics, consensus, outranking, tournament.

Introduction

When a group of people has an identical opinion concerning a set of decision alternatives, it makes no difference which minimally reasonable social choice rule is used in making the collective decision. This seems like a truism and in a way it is. Its validity hinges, however, on what we mean by opinion, reasonable rule and collective decision. If the individual opinions are expressed - as they commonly are in social choice theory - as complete and transitive preference relations over the alternatives and if the collective decision is also such a relation, then it is easy to envision rules which result in collective decisions that differ from the unanimously held opinion. One example is the rule that converts the unanimous preference ranking into its mirror image, *i.e.* if x is preferred to y by all individuals, then y is preferred to x in the collective decision. Such a rule is, however, downright bizarre and, thus, cannot be regarded as reasonable.

[*]The authors thank an anonymous referee for several constructive comments on an earlier version.

The collective decision may, however, differ from the unanimously held individual opinion in its structure. For example, the collective decisions may be sets rather than rankings. Or they may be rankings over just a subset of alternatives. With unanimous individual preference rankings, all reasonable choice rules preserve the relevant part of the individual ranking. If the winner is sought for, then the unanimously first-ranked alternative is the reasonable choice. If the best three alternatives should be found, then the three highest ranked alternatives in the collective ranking is the obvious choice. That the unanimously held opinion should be reflected in the social choice is one of the most obvious criteria one can impose on social choice rules.

But the setting where all individuals have an identical opinion on the alternatives is certainly not typical. If it were, there would not be much need for theorizing about social choices. It is easy to see that even a smallest deviation from perfect unanimity may bring about differences in choices under reasonable aggregation rules. Consider a profile with 5 voters and 6 alternatives so that 4 voters have an unanimous preference ranking $A \succ B \succ C \succ D \succ E \succ F$ and one voter has the ranking $B \succ C \succ D \succ E \succ F \succ A$ (Nurmi, 2005). Here the collective preference ranking ensuing from the Borda count differs from that held by all but one voter. This shows that the Borda count violates the no-veto condition which requires that whenever all voters save one agree on which alternative is best, then this alternative ought to be chosen (Maskin, 1985). Under the approval voting and assuming that all voters cast their vote for the same number of highest ranked alternatives, the nearly unanimously first-ranked A will not be collectively first-ranked unless all voters approve of just one alternative. All other acceptance thresholds result in some other alternative being collectively top-ranked.

In this paper we focus on choice rules that can be characterized in terms of a consensus profile and distance measure so that the application of the rule in effect amounts to minimizing the distance between consensus profile and the observed one. In effect, these methods look for the profile that is in some specific sense closest to the one observed and reflects in some relevant sense consensus among voters.

1.   The Methods

The best-known method of the distance-minimizing variety is Kemeny's rule (Kemeny, 1959). Given an observed preference profile, it determines the preference ranking over all alternatives that is closest to the observed one in the sense of requiring the minimum number of pairwise changes in individual opinions to reach that ranking. In the 5-voter example above, the closest collective ranking is the one held by all but one voter since this can be reached from the observed profile by making 5 preference changes (one pushing A above F, the second taking it above E, *etc.*). All other candidates for unanimous collective preference rankings require a strictly larger number of pairwise inversions of alternatives.

Kemeny's rule has many desirable properties as a choice rule: it is monotonic, always chooses a Condorcet winner if one exists (*i.e.* is a Condorcet extension), never chooses the eventual Condorcet loser, to name a couple. By the same token, it also has all those drawbacks that Condorcet extensions necessarily have. Notably, it is vulnerable to the no-show paradox, in fact, even to the strong version thereof (Pérez, 2001). This means that under Kemeny's rule it may happen that a voter is better off by not voting at all than voting according to his/her (hereinafter his) preferences. Indeed, he might get his first-ranked alternative elected if he refrains from voting, while it is not elected if he votes according to his preferences.

It has been suggested that Kemeny's rule is, in fact, what Marquis de Condorcet had in mind when devising his voting system in late 18'th century (Michaud, 1985; Young, 1988). Condorcet was preoccupied with maximizing the probability of reaching the correct collective decision, given that the individual members of the collectivity have a constant probability of being right. It can be shown (see Young (1988)) that the maximum likelihood consensus ranking is the one obtained by applying Kemeny's method.

The Borda count is typically mentioned in connection with positional preference aggregation systems.[1] Indeed, the usual definition of the system determines the Borda ranking on the basis of the positions that various alternatives occupy in individual preference rankings. If the number of alternatives is k, each first rank gives $k - 1$ points, each second rank $k - 2$ points *etc.* for each alternative. The Borda score of an alternative is the sum of points it receives from all voters. The Borda ranking, in turn, is the ranking determined by the magnitude of the Borda scores.

As Borda pointed out, the method can be implemented purely on the basis of pairwise comparisons. In the $k \times k$ matrix of comparisons, let the entry (i, j) denote the number voters who prefer alternative i to alternative j, with $(i, i) = 0$. Summing over row i of the matrix then gives us the Borda score of alternative i.

Writing about a century later than Borda and Condorcet, C.L. Dodgson proposed a voting system that - under specific circumstances - can be implemented on the basis of pairwise comparisons, but in general requires also positional information regarding voter preferences. The system works as follows. Given a

---

[1]The Borda count has apparently been proposed much before the time of Jean-Charles de Borda. McLean and Urken attribute this system to Ramon Lull who in somewhat ambiguous terms outlined it in a novel written in late 13'th century (McLean and Urken, 1995, 16-19). It is, however, fair to say that Borda was the first writer to systematically examine the system.

preference profile, determine first if there is an alternative that beats all the others in pairwise comparisons with a majority of votes. If such an alternative, the Condorcet-winner, exists, then it is the winner. Otherwise, that is, when there is no Condorcet winner, one tallies the number of pairwise preference changes that are needed to make each alternative the Condorcet winner. That alternative which requires the smallest number of such changes is the winner in Dodgson's sense. Obviously, Dodgson's procedure allows us to construct a social ranking over the alternatives: the fewer the changes that are needed to make an alternative the Condorcet winner, the higher its rank.

Copeland's system is of more recent origin although it has very likely been used for a long time under various names. It is based on pairwise comparisons (Copeland, 1951). In fact, one never needs more than this information to compute the Copeland winner and ranking. The basic version of the system tallies the number of alternatives that a given alternative defeats by a majority of votes. This number is the Copeland score of the alternative. The Copeland ranking is the same as the order of the Copeland scores: the higher the score, the higher the ranking.

Slater's rule combines features of Copeland with those of Kemeny (Slater, 1961). Given a preference profile, one constructs the corresponding tournament matrix, $i.e.$ a $0-1$ matrix where 1 in ith row and jth column indicates that ith alternative defeats the jth one by a majority. Otherwise the entry is 0. Let us call it the observed tournament matrix over the alternatives. This may or may not be cyclic. One then compares this matrix with all those tournaments that represent complete and transitive relations over the same alternatives. Finally, one computes the distance of the observed matrix with all those representing complete and transitive rankings. The distance is measured by the number of binary switches of preferences needed to transform the observed matrix into one representing a complete and transitive relation (Nurmi, 2002, 326-328).

Three more recent methods will also be dealt with in this paper: Litvak's, Tideman's and Schulze's (Litvak, 1982; Bury and Wagner, 2003; Tideman, 1987; Zavist and Tideman, 1989; Schulze 2003). Like Borda's also Litvak's method is based on a scoring rule. Each preference ranking is assigned a vector of k components. The component i indicates how many alternatives are placed ahead of the i'th one in the ranking under consideration. Each such vector thus consists of numbers $0, \ldots, k-1$. Litvak's rule looks for the k-component vector V that is closest to the observed preferences in the sense that the sum of component-wise absolute differences between the reported preference vectors and V is minimal.

Tideman's method is a sequential one proceeding from the largest pairwise victory of an alternative against another. Assume that x_1 and x_2 form such a pair. Denoting by v_i the number of votes x_i receives, this means that $v_1 - v_2$ is the largest difference of votes. We preserve this pair by drawing an arrow from x_1 to x_2. We then look for the next largest victory margin and preserve it similarly. Proceeding in this manner we may encounter a situation where preserving a pair would create a cycle. In these situations the respective relation between alternatives is ignored. Otherwise we proceed in decreasing order of victory margins. Eventually all alternatives are located in a directed graph with no cycles. The root of the graph is the winning alternative. Since by construction the cycles are excluded, Tideman's method yields a ranking over the set of alternatives.

Schulze's method, in turn, determines the "beatpaths" from each alternative x to every other alternative y. The path consists of ordered alternative pairs in the chain $x, w_1 \ldots, w_j, y$ - where $w_1, \ldots, w_j$ are alternatives - that leads from x to y. The strength of the beatpath is the smallest margin of victory in the chain. There are typically several beatpaths leading from one alternative to another. Each of them, thus, has the strength of its weakest link. Denote now by S_{xy} the strength of the strongest path from x to y. In other words, all but the strongest path between any two alternatives are ignored. Schulze's (potential) winner is the alternative x_S for which $S_{x_S y} \geq S_{y x_S}$ for all alternatives y.[2] Schulze proves that at least one such candidate always exists. Moreover, the relation $S_{xy} > S_{yx}$ is transitive. That is, if $S_{xy} > S_{yx}$ and $S_{yz} > S_{zy}$, then also $S_{xz} > S_{zx}$, for all alternatives x, y, z.

In addition to the above somewhat technical systems, we shall also deal with the plurality voting which is perhaps the best-known voting procedure. It is often regarded as the embodiment of the one-person-one-vote principle. In this system every individual has one vote and the winning alternative is one with the largest vote sum.

2.   Consensus States and Metrics

Apart from Kemeny's rule the above systems do not explicitly contain an idea that the voting outcome would be such a consensus state which is nearest to the reported preferences. Yet, upon closer scrutiny this idea can be associated with each of the systems outlined above. The state from which the distance to the observed profile in Kemeny's system is measured is one of unanimity regarding all positions of the ranking of alternatives, *i.e* the voters are in agreement about which alternative is placed first, which second *etc.* throughout all positions. The metric used in measuring the distance from the consensus is the inversion metric. Let us briefly remind ourselves of these concepts.

We define a distance function over a set P (of points, for example) as any function $d : P \times P \to R^+$, where R^+ is the set of non-negative real numbers.

---

[2]In case there are several potential winners, Schulze suggest a tie-breaking procedure which, however, will not be discussed here (Schulze, 2003, 3).

A distance function d_m is called a metric if the the following conditions are met for all elements x, y, z of P:

1 $d_m(x, x) = 0$,

2 if $x \neq y$, then $d_m(x, y) > 0$,

3 $d_m(x, y) = d_m(y, x)$, and

4 $d_m(x, z) \leq d_m(x, y) + d_m(y, z)$.

Substituting preference relations for elements in the above conditions we can extend the concept of distance function to preference relations. But these conditions leave open the way in which the distance between two relations is measured. Kemeny's proposal is the following (Kemeny (1959), see also Baigent (1987a,b)). Let R and R' be two rankings. Then their distance is:

$$d_K(R, R') = |\{(x, y) \in X^2 \mid R(x) > R(y), \ R'(y) > R'(x)\}|.$$

Here we denote by $R(x)$ the number of alternatives worse than x in a ranking R. This is called inversion metric.

The distance of two rankings is the number of inversions of consecutive choices needed to transform one ranking into another. This can be easily seen because:

- If two consecutive choices are in wrong order they must be inverted at some point

- If two consecutive choices are in right order they need not to be inverted.

The distance between two preference profiles of the same size is, then, the sum of the distances between the individual rankings. That is,

$$d_K(P, P') = \sum_{i=1}^{n} d_K(R_i, R'_i),$$

where the profile P consists of rankings $R_1, \ldots, R_n$ and the profile P' of rankings $R'_1, \ldots, R'_n$. Similarly, we can measure the distance between a profile P and a set of profiles $\mathbf{S}$,

$$d_K(P, \mathbf{S}) = \min_{P' \in \mathbf{S}} d_K(P, P').$$

We can generalize any distance function of rankings this way.

Let $U(R)$ denote an unanimous profile where every voter's ranking is R. Kemeny's rule results in the ranking $\bar{R}$ so that

$$d_K(P, U(\bar{R})) \leq d_K(P, U(R)) \quad \forall R \in \mathcal{R} \setminus \bar{R}$$

where P is the observed profile and $\mathcal{R}$ denotes the set of all possible rankings. If all the inequalities above are strict then $\bar{R}$ is the only winner.

To turn now to the Borda count, let us consider an observed profile P. For a candidate x we denote by $\mathbf{W}(x)$ the set of all profiles where x is first-ranked in every voter's ranking. Clearly in all these profiles x gets the maximum points. We consider these as the consensus states for the Borda count (Nitzan, 1981; Farkas and Nitzan, 1979).

For a candidate x, the number of alternatives above it in any ranking of P equals the number of points deducted from the maximum points. This is also the number of inversions needed to get x in the winning position in every ranking. Thus, using the metric above, w_B is the Borda winner if

$$d_K(P, \mathbf{W}(w_B)) \leq d_K(P, \mathbf{W}(x)) \quad \forall x \in X \setminus w_B.$$

The plurality system is obviously also directed at the same consensus state as the Borda count, but its metric is different. Rather than counting the number of pairwise preference changes needed to make a given alternative unanimously first ranked, it minimizes the number of individuals having different alternatives ranked first.

We define a discrete metric as follows

$$d_d(R, R') = \begin{cases} 0 & \text{if } R = R', \\ 1 & \text{otherwise.} \end{cases}$$

As above the distance between two preference profiles of the same size is

$$d_d(P, P') = \sum_{i=1}^{n} d_d(R_i, R_i'),$$

where the profile P consists of rankings $R_1, \ldots, R_n$ and the profile P' of rankings $R_1', \ldots, R_n'$. That is, the distance indicates the number of rankings that differ in the two profiles. Note that we consider the profiles always in such a way that the order of the rankings in the profiles is irrelevant. Also, the distance between a profile P and a set of profiles $\mathbf{S}$ is, similarly,

$$d_d(P, \mathbf{S}) = \min_{P' \in \mathbf{S}} d_d(P, P').$$

The unanimous consensus state in plurality voting is one where all voters have the same alternative ranked first. With the metric, in turn, we tally for each alternative, how many voters in the observed profile do not have this alternative as their first ranked one. The alternative for which this number is smallest is the plurality winner. The plurality ranking coincides with the order of these numbers.

Using this metric we have for the plurality winner w_p,

$$d_d(P, \mathbf{W}(w_p)) \leq d_d(P, \mathbf{W}(x)) \quad \forall x \in X \setminus w_p.$$

The only difference to the Borda winner is the different metric used.

Dodgson's system is based on a different idea of the goal state, *viz.* one where there is a Condorcet winner. For any candidate x we denote by $\mathbf{C}(x)$ the set of all profiles where x is the Condorcet winner.

Provided that a Condorcet winner exists in the observed profile, the winner is this alternative. Otherwise, one constructs, for each alternative x, a profile in $\mathbf{C}(x)$ which is obtained from the observed profile P by moving x up in one or several voters' preference orders so that x emerges as the Condorcet winner. Obviously, any alternative can thus be rendered a Condorcet winner. It is also clear that for each alternative there is a minimum number of such preference changes involving the improvement of x's position *vis-à-vis* other alternatives that are needed to make x the Condorcet winner. The Dodgson winner w_D is then the alternative which is closest, in the sense of Kemeny's metric, of being the Condorcet winner. That is,

$$d_K(P, \mathbf{C}(w_D)) \leq d_K(P, \mathbf{C}(x)) \quad \forall x \in X \setminus w_D.$$

Dodgson's method is thus characterized by Kemeny's inversion metric combined with a goal state where a Condorcet winner exists.

Litvak's procedure results in the ranking that is nearest to the observed one in terms of minimizing the sum (over individuals) of absolute rank position differences of alternatives in the former and the latter. As the position numbers play an important role in the Borda count, one would expect that Litvak's method is similar to the Borda count. This is, however, not the case (Nurmi, 2004, 8-9). For example, Litvak's procedure may end up with a Condorcet loser being ranked first which is never the case under the Borda count. Litvak's procedure also differs for Kemeny's. To see this, consider the distance of $A \succ B \succ C \succ D$ to $C \succ D \succ A \succ B$, on the one hand, and to $D \succ C \succ B \succ A$, on the other. In Kemeny's sense, the latter difference is larger than the former, while in Litvak's sense they are equidistant from $A \succ B \succ C \succ D$.

We define the Litvak metric formally as follows:

$$d_L(R, R') = \sum_{x \in X} |R(x) - R'(x)|.$$

The Litvak winning ranking R_L has the property

$$d_L(P, U(R_L)) \leq d_L(P, U(R)) \quad \forall R \in \mathcal{R} \setminus R_L$$

exactly like the Kemeny winner except for the different metric.

The Litvak and Kemeny metrics are related in the following way:

- If a ranking R' is derived from ranking R by only moving one candidate up (or down) by n steps then $2d_K(R, R') = 2n = d_L(R, R')$

- If a ranking R cannot be turned into ranking R' by switching adjacent candidates without moving at least one candidate at some point first up then down (or first down then up), then $d_K(R, R') < d_L(R, R') < 2d_K(R, R')$

An example: reversing order $A \succ B \succ C$ this way requires moving B to both directions. The distance of that order and its reverse is 3 using Kemeny metric and 4 using Litvak metric: $3 < 4 < 6$.

Because of the first property we have that for any P and x

$$d_L(P, \mathbf{W}(x)) = 2d_K(P, \mathbf{W}(x))$$

and

$$d_L(P, \mathbf{C}(x)) = 2d_K(P, \mathbf{C}(x)).$$

Thus we find the same Borda count and Dodgson winners using either of the metrics.

For the sake completeness we shortly consider what we get if we combine the discrete metric with consensus states $U(R)$ or $\mathbf{C}(x)$. In the former case simply the most popular ranking is selected; *that is, the ranking with least opposition.*

The latter case is more interesting. We can formally define the winner w_Y as the option with property

$$d_d(P, \mathbf{C}(w_Y)) \leq d_d(P, \mathbf{C}(x)) \quad \forall x \in X \setminus w_Y.$$

In other words we find the largest set of voters such that the Condorcet winner exists. This system has been attributed to H. P. Young (Smith, 2005).

3.   Outranking and Tournament Matrices

We now move to different kind of systems. In the following we define the distance of two profiles, not rankings. Also the distance is calculated using outranking matrices instead of individual rankings.

We denote by V the outranking matrix where entry V_{xy} indicates the number of voters in profile P preferring candidate x to candidate y. The diagonal entries are left blank.

We define the metric using outranking matrices as follows: if V and V' are the outranking matrices of profiles P and P' then

$$d_V(P, P') = \frac{1}{2} \sum_{x,y \in X} |V_{xy} - V'_{xy}|.$$

In other words, the distance tells us how much pairwise comparisons differ in the corresponding outranking matrices.

This metric is very similar to the inversion metric and, indeed, we find the same Borda count (or, alternatively, Kemeny) winners using this metric instead.

The interesting case is the one where the goal is a Condorcet winner.

In the Condorcet least-reversal system the winner is the candidate which can be turned into Condorcet winner with minimum number of reversals of pairwise comparisons. That is, the Condorcet least-reversal system winner w_{lr} is the candidate with property

$$d_V(P, \mathbf{C}(w_{lr})) \leq d_V(P, \mathbf{C}(x)) \quad \forall x \in X \setminus w_{lr}.$$

Copeland's procedure has also a similar goal state as Dodgson's and Condorcet's least reversal one, namely, one with a Condorcet winner. Given an observed profile P one considers the corresponding tournament matrix T where entry $T_{xy} = 1$ if majority of the voters in profile P are preferring candidate x to candidate y. Otherwise $T_{xy} = 0$.

Now, the Condorcet winner is seen as a row in T where all $k - 1$ non-diagonal entries are 1s.

We define yet another distance between profiles P and P' with tournament matrices T and T' as

$$d_T(P, P') = \frac{1}{2} \sum_{x,y \in X} |T_{xy} - T'_{xy}|.$$

The Copeland winner w_C is the alternative that wins the largest number of comparisons with other candidates i.e. has the smallest number of 0s in its row in the tournament matrix. Thus the winner w_C is the candidate that comes closest to win every other candidate, that is, using the distance above,

$$d_T(P, \mathbf{C}(w_C)) \leq d_T(P, \mathbf{C}(x)) \quad \forall x \in X \setminus w_C.$$

Obviously, the goal states of Condorcet least-reversal system and Copeland's system are the same, but metrics differ. The latter pays no attention to majority margins, while the former depends on them.[3]

We can expand the idea of the Condorcet winner into a Condorcet ordering. Let P be a profile and X' be a subset of the set of candidates X. We denote by $P|_{X'}$ the profile we get when we only consider the candidates in X' and dismiss all other candidates. We say that a profile P has a Condorcet ordering if for every $X' \subset X$ the profile $P|_{X'}$ has a Condorcet winner.

---

[3] See, Klamler (2005) for another distance based characterization of the Copeland rule.

More precisely, there is a candidate $w_1 \in X$ which wins every other candidate in pairwise comparisons with a majority of votes. And there is a candidate w_2 which wins every other candidate except w_1 in pairwise comparisons with a majority of votes. And there is a candidate w_3 which wins every other candidate except w_1 and w_2 in pairwise comparisons with a majority of votes. And so on. The Condorcet ordering is $w_1 \succ w_2 \succ \cdots \succ w_k$, where $k = |X|$.

It can be easily seen that P has a Condorcet ordering if and only if the tournament matrix of P does not have any cycles, i.e. there are no candidates $x_1, x_2, \ldots, x_i$ such that $T_{x_1 x_2} = T_{x_2 x_3} = \cdots = T_{x_{i-1} x_i} = T_{x_i x_1} = 1$.

We denote by $\mathbf{Co}(R)$ the set of all profiles that have Condorcet ordering R. This set is a natural goal for a decision rule. But, as above there are several different methods to decide which ordering is closest to the profile that does not have a Condorcet ordering.

Two natural measures are the number of pairwise comparisons that need to be reversed and the number of pairwise losses with majority that need to be turned into wins. That is, we could measure the differences in either outranking or tournament matrices. The latter of these methods is connected to the work of P. Slater (Slater, 1961),

$$d_T(P, \mathbf{Co}(R_{Sl})) \le d_T(P, \mathbf{Co}(R)) \quad \forall R \in \mathcal{R} \setminus R_{Sl}.$$

The other method does not have a name:

$$d_V(P, \mathbf{Co}(R_{U_1})) \le d_V(P, \mathbf{Co}(R)) \quad \forall R \in \mathcal{R} \setminus R_{U_1}.$$

The other three winners we can define with the metrics mentioned above are also yet to be named:

$$d_K(P, \mathbf{Co}(R_{U_2})) \le d_K(P, \mathbf{Co}(R)) \quad \forall R \in \mathcal{R} \setminus R_{U_2}$$

$$d_L(P, \mathbf{Co}(R_{U_3})) \le d_L(P, \mathbf{Co}(R)) \quad \forall R \in \mathcal{R} \setminus R_{U_3}$$

$$d_d(P, \mathbf{Co}(R_{U_4})) \le d_d(P, \mathbf{Co}(R)) \quad \forall R \in \mathcal{R} \setminus R_{U_4}.$$

4.   Metrics Based on the Elimination of Candidates

Above we have defined the metric counting the differences in the individual rankings (d_K, d_L, d_V), the pairwise defeats (d_T), and the inconsistent rankings (d_d). Another approach would be to consider the candidates that vary in two preference profiles.

There are several recursive elimination systems that are based on the order of the eliminations of the candidates.

Let R be some ranking of the candidates, $x_1 \prec x_2 \prec \cdots \prec x_k$ and P the voting profile we are considering. We begin by comparing the Hare system and the function

$$F_P(R) = \sum_{i=1}^{k}(n + 1 - d_d(P_{i-1}, \mathbf{W}(x_i)))(n + 1)^{k-i}$$

where $P_{i-1} = P|_{X \setminus \bigcup_{j=1}^{i-1} x_j}$.

On the first round of the Hare system we eliminate the candidate with the smallest number of first places. The function F_P gets its smallest value when the function

$$F'_P(x_1) = (n + 1 - d_d(P, \mathbf{W}(x_1)))(n + 1)^{k-1}$$

gets its smallest value. This happens when x_1 is the candidate with the smallest number of first places in P and thus $d_d(P, \mathbf{W}(x_1))$ is maximal. Note that the part

$$\sum_{i=2}^{k}(n + 1 - d_d(P_{i-1}, \mathbf{W}(x_i)))(n + 1)^{k-i}$$

of $F_P(R)$ is always smaller than $(n + 1)^{k-1}$. That is why we do not need to care about candidates $x_2, x_3, \ldots, x_k$ at this point.

On the second round of the Hare system we eliminate the candidate with the smallest number of first places after we have removed the candidate that was eliminated at the previous round. The function F_P gets its smallest value when the function

$$F''_P(x_1, x_2) = (n + 1 - d_d(P, \mathbf{W}(x_1)))(n + 1)^{k-1} \\ + (n + 1 - d_d(P|_{X \setminus x_1}, \mathbf{W}(x_2)))(n + 1)^{k-2}$$

gets its smallest value. This happens when x_1 is the candidate with the smallest number of first places and x_2 is the candidate with the smallest number of first places after x_1 is eliminated from P. Note again that the part

$$\sum_{i=3}^{k}(n + 1 - d_d(P_{i-1}, \mathbf{W}(x_i)))(n + 1)^{k-i}$$

of $F_P(R)$ is always smaller than $(n + 1)^{k-2}$. That is why we do not need to care about candidates $x_3, x_4, \ldots, x_k$ at this point.

If we continue these considerations we obtain that the function F_P gets its smallest value when $x_1, x_2, \ldots$ is the order of the eliminations using the Hare system.

Next we turn this function into a metric. A metric is always symmetric. Let P and P' be two preference profiles of n rankings and their sets of candidates

X and X' have k and k' elements, respectively. We write shortly $P_{i-1} = P|_{X \setminus \bigcup_{j=1}^{i-1} x_j}$. Let the metric for the Hare system be

$$d_H(P, P') = \min \left\{ \sum_{i=1}^{m} (n + 1 - d_d(P_{i-1}, \mathbf{W}(x_i)))(n+1)^{k-i} + \right.$$

$$\left. \sum_{i=1}^{m'} (n + 1 - d_d(P'_{i-1}, \mathbf{W}(x'_i)))(n+1)^{k'-i} \mid x_i \in X, x'_i \in X', P_m = P'_{m'} \right\}.$$

Because $d_H(P, P)$ must be zero for any P, we only calculate the distance until $P_m = P'_{m'}$ for some m and m'.

Note that when we calculate the distance between a profile P and the sets $\mathbf{W}(x)$ (or, alternatively, $\mathbf{C}(x)$) there is always a profile in the closest set $\mathbf{W}(x)$ (or $\mathbf{C}(x)$) such that the second part of the distance function is zero.

When we combine this metric with goal states $\mathbf{W}(x)$ we get the Hare system. If we instead use the goal states $\mathbf{C}(x)$ we get a system named after Thomas Hill. For the Coombs method we have the metric

$$d_C(P, P') = \min \left\{ \sum_{i=1}^{m} (d_d(P_{i-1}, \mathbf{L}(x_i)) + 1)(n+1)^{k-i} \right.$$

$$\left. + \sum_{i=1}^{m'} (d_d(P'_{i-1}, \mathbf{L}(x'_i)) + 1)(n+1)^{k'-i} \mid x_i \in X, x'_i \in X', P_m = P'_{m'} \right\},$$

where we denote by $\mathbf{L}(x)$ the set of all profiles where x is last-ranked in every voter's ranking Finally, for the Baldwin, also called Borda runoff, method we have

$$d_B(P, P') = \min \left\{ \sum_{i=1}^{m} (kn - d_K(P_{i-1}, \mathbf{W}(x_i)))(kn)^{k-i} \right.$$

$$\left. + \sum_{i=1}^{m'} (k'n - d_K(P'_{i-1}, \mathbf{W}(x'_i)))(k'n)^{k'-i} \mid x_i \in X, x'_i \in X', P_m = P'_{m'} \right\}.$$

5.   Two More Systems

Turning now to somewhat more recent systems, Tideman's procedure operates on pairwise majority margins. At every step the algorithm fixes one comparison between two candidates such that a minimum amount of voters are disappointed. The result is a complete directed graph without cycles i.e. a ranking. Clearly the idea of this procedure is to generate a ranking that contradicts as few pairwise comparisons as possible in the outranking matrix. This is exactly what the Kemeny rule does. While Kemeny's always chooses the ranking that is closest to the profile Tideman's procedure *tries* to find that ranking

Table 1. Goal states, metrics and voting systems.

Goal state	Unanimous		Condorcet		Beatpath
Metric	winner $\mathbf{W}(x)$	order $U(R)$	winner $\mathbf{C}(x)$	order $\mathbf{Co}(R)$	winner/ order
Inversion, d_K	Borda	Kemeny	Dodgson	U 2	Schulze
Manhattan, d_L	Borda	Litvak	Dodgson	U 3	Schulze
Inversion for V matrices, d_V	Borda	Kemeny	Condorcet least-reversal	U 1	Schulze
Inversion for T matrices, d_T	-	$-^4$	Copeland	Slater	Schulze
Discrete, d_d	Plurality	Plurality	Young	U 4	Schulze
Instant runoff, d_H	Hare	(Hare)	Hill	$-^4$	Schulze
Disapproval runoff, d_C	Coombs	(Coombs)[5]	U5	$-^4$	Schulze
Borda runoff, d_B	Baldwin	(Baldwin)[5]	Baldwin	$-^4$	Schulze

[4] The goal states and the metric are incompatible or their meaning is unclear.

[5] The reversed order of the eliminations can be interpreted as the resulting order.

using *greedy* algorithm. From the computational point of view the Tideman winner is always fast to find while finding the Kemeny winner can be very slow if the number of candidates is large.

The last system we consider is Schulze's method. Let us denote by S_{xy} the maximum strength of all beatpaths from x to y. Using these we generate the "beatpath tournament matrix" B of the profile P such that the entry $B_{xy} = 1$ if $S_{xy} \geq S_{yx}$; otherwise $B_{xy} = 0$. We say that the ranking R where $R(x) > R(y)$ iff $B_{xy} = 1$ corresponds to this beatpath tournament matrix.

As above we could define the Schulze winning ranking R_{Sc} in profile P with the help of a profile P_{Sc} that is closest to P (with respect to some metric d) and has a beatpath tournament matrix that corresponds to some complete ordering R_{Sc}: Let P_{Sc} be a beatpath tournament matrix that corresponds to complete ordering R_{Sc} and

$$d(P, P_{Sc}) \leq d(P, P')$$

for all profiles P' that have a beatpath tournament matrix that corresponds to some complete ordering.

But every profile has a beatpath tournament matrix that corresponds to complete ordering. Thus, regardless of the metric, $d(P, P_{Sc}) = 0$ and the winning ranking R_{Sc} is the ordering corresponding to the beatpath tournament matrix of P.

6.   Conclusion

The observations made in the preceding are summarized in Table 1.

References

Baigent, N. (1987a) Preference proximity and anonymous social choice. *The Quarterly Journal of Economics 102*, 161-169.

Baigent, N. (1987b) Metric rationalization of social choice functions according to principles of social choice. *Mathematical Social Sciences 13*, 59-65.

Bury, H. and Wagner, D. (2003) Use of preference vectors in group judgement: the median of Litvak. In Kacprzyk, J. and Wagner, D. (eds), *Group Decisions and Voting*. Exit, Warszawa.

Copeland, A. H. (1951) A 'reasonable' social welfare function. Mimeo. University of Michigan, Seminar on Applications of Mathematics to the Social Sciences. Ann Arbor.

Farkas, D. and Nitzan, S. (1979) The Borda rule and Pareto stability: A comment. *Econometrica 47*, 1305-1306.

Kemeny, J. (1959) Mathematics without numbers. *Daedalus 88*, 571-591.

Klamler, Chr. (2005) Copeland's rule and Condorcet's principle. *Economic Theory 25*, 745-749.

Litvak, B. G. (1982), *Information Given by the Experts. Methods of Acquisition and Analysis*. Radio and Communication, Moscow (in Russian).

Maskin, E.S. (1985) The theory of implementation in Nash equilibrium. In: Hurwicz, L., Schmeidler, D., Sonnenschein, H. (eds) *Social Goals and Social Organization: Essays in Memory of Elisha Pazner*. Cambridge University Press, Cambridge.

McLean, I. and Urken, A. B. (1995), General introduction. In: McLean, I. and Urken, A.B. (eds) *Classics of Social Choice*. The University of Michigan Press, Ann Arbor.

Michaud, P. (1985) Hommage a Condorcet. *Centre Scientifique IBM France*, Report No F-094, November.

Nitzan, S. (1981) Some measures of closeness to unanimity and their implications, *Theory and Decision 13*, 129-138.

Nurmi, H. (2002) Measuring disagreement in group choice settings. In: Holler, M. J., Kliemt, H., Schmidtchen and Streit, M. E. (eds) *Power and Fairness*. Jahrbuch für Neue Politische Ökonomie, Band 20. Mohr Siebeck, Tübingen.

Nurmi, H. (2004) A comparison of some distance-based choice rules in ranking environments. *Theory and Decision 57*, 5-24.

Nurmi, H. (2005) A responsive voting system. *Economics of Governance 6*, 63-74.

Pérez, J. (2001) The strong no show paradoxes are common flaw in Condorcet voting correspondences. *Social Choice and Welfare 18*, 601-616.

Schulze, M. (2003) A new monotonic and clone-independent single-winner election method. *Voting Matters 17*, 9-19.

Slater, P. (1961) Inconsistencies in a schedule of paired comparisons. *Biometrika 48*, 303-312.

Smith, W. D. (2005) Descriptions of voting systems. Typescript.

Tideman, N. (1987) Independence of clones as a criterion for voting rules. *Social Choice and Welfare 4*, 185-206.

Young, H. P. (1988) Condorcet's theory of voting. *American Political Science Review 82*, 1231-1244.

Zavist, B. T. and Tideman, N. (1989) Complete independence of clones in the ranked pairs rule. *Social Choice and Welfare 6*, 167-173.

A Strategic Problem in Approval Voting

Jack H. Nagel*

Political Science Department, University of Pennsylvania

Abstract   Problems of multi-candidate races in U.S. presidential elections–exemplified by Ralph Nader's spoiler effect in 2000–motivated the modern invention and advocacy of approval voting; but it has not previously been recognized that the first four U.S. presidential elections (1788-1800) were conducted using a variant of approval voting. That experiment ended disastrously in 1800 with an infamous Electoral College tie between Thomas Jefferson and Aaron Burr. The tie, this paper shows, resulted less from miscalculation than from a strategic tension built into approval voting, which forces two leaders appealing to the same voters to play a game of Chicken. All outcomes are possible, but none is satisfactory–mutual cooperation produces a tie, while all-out competition degrades the system to single-vote plurality, which approval voting was designed to replace. In between are two Nash equilibria that give the advantage to whichever candidate enjoys an initial lead or, in the case of initial parity, to the candidate who is less cooperative and more treacherous.

Keywords: approval voting, U.S. presidential elections, Aaron Burr, election of 1800, electoral systems, voting methods

1.   Introduction

Approval voting is a balloting method that allows each elector to vote for, or "approve", as many alternatives as he or she desires, even if that number exceeds the number to be chosen. In contrast to preferential (rank-order) ballots, each vote counts equally. Invented (or rediscovered) independently by three groups of social scientists in the 1970s, approval voting is designed chiefly to solve problems that arise when three or more candidates compete in single-winner, plurality rule (first-past-the-post) elections. Foremost among these problems is the danger that a minority will prevail if the majority divides its votes between two candidates who appeal to the same voters.

---

*For helpful, if sometimes dissenting, comments, I am grateful to Samuel Merrill, Robert Norman, and participants in the Erice workshop, especially Steven Brams.

American advocates of approval voting are motivated mainly by concern about U.S. presidential elections–both the crowded fields of candidates typical in early primaries and the frequent presence of significant third-party contenders in general elections. It is therefore surprising that no one has previously noticed that a variant of approval voting was used in the first four U.S. presidential elections–1788, 1792, 1796, and 1800. Unfortunately, that early experience was inauspicious. The 1800 election produced an Electoral College tie between Thomas Jefferson and his running mate, Aaron Burr. As the Constitution provided (and still does today), the decision then passed to the House of Representatives. Voting in February 1801, the House remained deadlocked for thirty-five ballots. On the thirty-sixth ballot, Jefferson was elected president, thus pulling the fledgling democracy back from a constitutional crisis and the brink of civil war. Before the approval-voting feature could be used again, it was repealed in 1804 by the Twelfth Amendment to the U.S. Constitution, which substituted a conventional balloting system that gives each elector only one vote per office.

Because the connection to approval voting has been overlooked, the flaw in the original presidential election system is usually viewed as a mere historical oddity, due to the Framers' inability to foresee the emergence of contests dominated by political parties. Similarly, the debacle of 1800-01 is remembered as a curious accident, owing to a failure of coordination within the victorious Republican Party of Jefferson and Burr.[1] Instead, this paper will argue that the outcome of 1800 resulted not from an accidental mistake, but rather from a strategic dilemma inherent in the original method of voting for president. I shall call that problem the Burr Dilemma after a man whom history has seldom honored in any other way.

The thesis of this chapter is that the Burr Dilemma remains a potential problem for modern approval voting, though less obviously than in the system of 1788-1800. An initial historical section describes the original U.S. presidential election system, relates it to approval voting, and shows how a strategic problem emerged in the first three elections and then caused a debacle in 1800. The second section defines the Burr Dilemma, contends that the early U.S. voting method is relevant to modern approval voting, analyzes the dilemma as a game of Chicken, and discusses how it can pose a problem for competitors under approval voting just as it did for the candidates of 1800.

---

[1] Following the lead of most historians, I use the original name for the party organized by Jefferson and James Madison. They called themselves "Republicans" to signal opposition to the alleged monarchical tendencies of the governing Federalists, led by Alexander Hamilton and John Adams. The label soon gave way to "Democratic-Republicans" and then to "Democrats," a term originally used by their adversaries as a pejorative.

2.   Approval Voting in the First Four Presidential Elections

Hostile to "mob rule" and the "mischief of faction," the Framers of the U.S. Constitution entrusted election of the President not to the people directly, but to the Electoral College, intending it to be an elite group capable of making enlightened judgments. They conceived of voting by the electors in the spirit of the Condorcet jury theorem: not as a test of strength among competing interests, but rather as a collective method of reaching the correct decision about who was best suited to lead the government. Believing that the most likely temptation away from that ideal would arise from local interests and loyalties, the Framers devised approval voting as an antidote: Electors were to "vote by ballot for two persons, of whom one at least shall not be an inhabitant of the same State with themselves."[2]

Thus at least one vote would be cast for a man of "continental character," rather than a local favorite son. The candidate receiving the most votes, "if such number be a majority of the whole number of electors appointed," would become president. The runner-up would become vice president, having been identified by the collective judgment as the person next-best-qualified to be president. The office of vice president was created in part to ensure that electors would cast two votes, rather than leave the second vote blank.[3]

In the original system, each of the electors' two votes counted equally and both were tallied as if cast for the presidency. Thus the method of 1788-1800 embodied the essential feature of approval voting as invented in the late twentieth century–more than one equal vote could be cast for a single office. There were, however, four differences from modern approval voting: (a) Presidential electors could not cast more than two votes, whereas approval voting permits voters to approve more than two candidates if they wish. (b) Although not distinguished on the ballot, a second office (the vice presidency) was at stake, so in one sense, electors cast a number of votes equal to the number of candidates to be elected. (c) To win, a presidential candidate, then as now, had to receive votes from a majority of electors, whereas approval voting is usually recommended in combination with the plurality decision rule (Merrill and Nagel 1987). (d) The number of electors was small (ranging from 69 in 1788 to 138 in 1796 and 1800), whereas proponents of approval voting recommend it for both small-scale and mass elections. Each of those differences can be significant in its own way, but none affects the central nature of the Burr Dilemma and its relevance to approval voting today. I will justify this claim later, after

---

[2] U.S. Constitution, Article II, Section 1.3.
[3] This paragraph draws on Ackerman and Fontana 2004 and Ceaser 1979 (ch. 1).

defining the Burr Dilemma, but first it is necessary to summarize the remarkable events of 1800-01.

2.1   The Election of 1800

After the 1796 election resulted in the victory of John Adams over Thomas Jefferson by the narrow electoral vote margin of 71-68, the two leaders, now serving as president and vice president, prepared to compete again in 1800.[4] Correctly calculating that the votes of New York (which Adams carried in 1796) would be critical to Jefferson's chances, the Republicans in May 1800 designated Aaron Burr of that state as their choice for vice president. The preceding month, by a brilliant feat of electioneering in New York City, Burr had captured control of the state legislature for the Republicans. The legislature was to choose New York's twelve presidential electors using a unit rule system that would give all electors to the majority party.[5] To secure New York's votes for himself, the Virginian Jefferson needed a New Yorker as running mate, and Burr was an obvious choice.

In December, when the electors voted in their respective state capitals (as the Constitution requires), it became known that the deal between Jefferson and Burr had worked all too well–the two Republicans were tied with 73 votes each, three above the 70-vote majority required for election. Adams followed with 65. His Federalist Party avoided a tie of their own because one Rhode Island elector "threw away" his second vote to John Jay, thus "cutting" the Federalist vice presidential candidate, Charles Cotesworth Pinckney of South Carolina, who received 64 votes.

To break a tie such as the one between Jefferson and Burr, the Constitution provides that the House of Representatives must choose between the top two candidates, with each state's delegation casting just one vote and a majority of all the states required to elect a winner. At the time, the union consisted of sixteen states, so the votes of nine were needed to resolve the deadlock. Republicans controlled the delegations of only eight states. Federalists dominated six caucuses; and two states (Vermont and Maryland) were evenly divided, so they could not vote as long as members of the two parties cancelled each other. The Federalists decided to support Burr, thus exploiting their blocking power to prevent election of their arch-rival, Jefferson.

---

[4] All electoral votes are from the website of the U.S. National Archives and Records Administration, http://www.archives.gov/federal_register/electoral_college/votes. My account of early elections draws on van der Linden 1962, Daniels 1970, Freeman 2002, Randall 1993, McCullough 2001, and Wills 2003.

[5] As many people learned to their dismay during the 2000 Florida fiasco, the Constitution gives each state's legislature the power to decide how its electors will be chosen (Article II, Section 1.2). In 1800, most legislatures retained for themselves the power to appoint electors. In the early years of the Republic, states moved to the unit rule in a competitive effort to gain, or avoid loss of, relative voting power. In the election of 1800, only three states chose divided delegations of electors.

What did Federalists hope to gain by creating a deadlock and thus a constitutional crisis? Three possibilities enticed them: (1) If the impasse continued long enough, their party might retain control of the presidency by extra-constitutional maneuvers, such as putting the president pro tempore of the Senate in charge of the executive branch.[6] (2) Alternatively, a handful of Republicans in three states might eventually shift their votes to Burr, thus electing the Republican many Federalists considered more pliable or, because he was a Northerner, ideologically more compatible.[7] (3) As a last resort, the Federalists might use their position to extract vital policy concessions before allowing either Republican to win. Not all Federalists approved of their party's strategy. In particular, Alexander Hamilton bombarded Federalist Congressmen with confidential letters arguing that his longtime foe Jefferson would be less dangerous than Burr, whom he condemned as a man of "no principle, public or private....sanguine enough to hope every thing, daring enough to attempt every thing, wicked enough to scruple nothing." (van der Linden 1962, 262)

On February 11, 1801, the House assembled in snowbound Washington, the raw new capital. Members vowed to remain in session until a president was chosen. Through six days and 35 ballots, the two sides held firm. Jefferson's fate was in the hands of Federalist Representative James Bayard. As tiny Delaware's sole member of the House, Bayard controlled his state's pivotal vote. He was also the acknowledged leader of the five Federalists from Maryland and Vermont, any of whom could elect Jefferson simply by abstaining. Bayard was willing to listen to Hamilton, whom he admired; but he continued to vote for Burr while seeking from either Jefferson or Burr a signal of policy concessions on four "cardinal points:" preservation of the Federalists' fiscal system, neutrality between England and France, expansion of the navy, and retention of Federalist officeholders (van der Linden 1962, 306). Suspicion and antagonism on both sides grew more and more fevered. Encouraged by Jefferson, the Republican governors of Pennsylvania and Virginia, Thomas McKean and James Monroe (a future president), prepared to mobilize their states' militias so they could resist with force any "usurpation" by Burr and the Federalists (Wills 2003, 84-6).

To Bayard's consternation, Burr refused to deal. Although some of Burr's friends lobbied Congress to promote his election, the candidate absented him-

---

[6]Although the administration of George Washington (1789-97) was initially pre-party and remained ostensibly above party, Federalists had effective control of the executive throughout the first twelve years of the new constitutional system. In transferring power from one party to another as a result of electoral processes, the events of 1800-01 established a precedent of enormous importance for democracy in the United States and around the world–but a peaceful transition very nearly failed to occur because of the tie vote.

[7]Jefferson owned slaves; and all of his Electoral College support came from Southern slave states, except for the twelve electors from New York and eight from a divided Pennsylvania delegation. Only ten of Adams' 65 electoral votes came from slave states.

self in Albany, preparing for the wedding of his beloved daughter Theodosia. Meanwhile, Bayard later charged, Jefferson was using promises of patronage appointments to prevent defections by Republicans from New York, New Jersey, Maryland, and Vermont. "Every man on whose vote the event of the election hung has since been distinguished by presidential favor," Bayard said in 1806, naming names (van der Linden 1962, 330). With no sign of any shift to Burr, Bayard finally received–so he thought–the policy assurances he wanted from Jefferson. According to Bayard, Republican Congressman Samuel Smith, who lived in the same boardinghouse as Jefferson, told him "that he had seen Mr. Jefferson, and stated to him the points mentioned, and was authorized by him to say that they corresponded with his views and intentions..."[8] Thus satisfied by Bayard, Federalists from Vermont and Maryland abstained on the thirty-sixth ballot, swinging their states' previously deadlocked votes to Jefferson, who thereby became the third president of the United States.[9]

2.2   The Strategic Problem of 1800 in Light of Earlier Elections

At first glance, it might seem that the tie between Jefferson and Burr was simply a failure of coordination, one that was not altogether surprising, given the Constitutional requirement that electors meet separately in their respective state capitals, which were hundreds of miles apart at a time of primitive

---

[8]When Bayard testified to this effect in 1806, Jefferson called his claim "absolutely false," offering a lawyerly denial (worthy of his namesake, William Jefferson Clinton) that hinged on the distinction between a "conversation" with Smith as opposed to "assurances to anybody...[about] what I would or would not do." (van der Linden 1962, 307)

[9]As is well known, hostilities inflamed by this crisis led to personal tragedies for two of the principal players (Daniels 1970). Aaron Burr had conducted himself with perfect propriety from a constitutional standpoint. Formally, the electoral votes he received were for president just as much as Jefferson's, and the final decision was for the House to make. Denying any wish to compete with Jefferson, he refrained from bargaining with the Federalists and generally kept himself above the fray. Burr refused, however, to accede to Republicans' entreaties that he pledge to renounce the presidency if he should win the vote in the House. Thus, from the partisan perspective of Jefferson and his allies, he became a betrayer and a would-be usurper. After Burr took office as vice president, Jefferson shunned him. In 1804, having been informed by Jefferson that he would not be renominated as vice president, Burr ran unsuccessfully for governor of New York. His enemy Hamilton again intervened with denunciations of Burr's character, some of which were reported in the press. Burr demanded that Hamilton deny the accusations or give him the satisfaction of a duel. Hamilton accepted the challenge. On July 11, 1804, at Weehawkin, New Jersey, Burr fatally shot his enemy, the first Secretary of the Treasury and co-author of the *Federalist Papers*. A fugitive from murder charges in both New Jersey and New York, Vice President Burr escaped to Washington, where he resumed his duties as presiding officer of the Senate, despite the scorn of most members. After his term ended, Burr's fortunes fell even lower. When he took a journey along the Ohio and Mississippi rivers, it was alleged that he conspired to lead a secession of western U.S. territories into a new empire under his own rule. Tried repeatedly at Jefferson's insistence for a variety of offenses, Burr was found not guilty by a series of juries. Afflicted by debts and public opprobrium, the former vice president spent four years in European exile before returning in 1812 to the U.S., where he lived the rest of his life in poverty and relative obscurity.

communication. As Jefferson wrote to Burr in December 1800, "It was badly managed not to have arranged with certainty what seems to have been left to hazard [i.e., the cutting of at least one vote from Burr]." (van der Linden 1962, 246) Closer examination reveals, however, that, far from being a mere oversight arising from excessive partisan solidarity, the tie was a logical outcome of the strategic dilemma created by the voting system, which exacerbated deep distrust between the two Republican candidates and the factions they led (Freeman 2002).

With two offices at stake and two undifferentiated votes at the disposal of each elector, candidates backed by the same party faced a fundamental tension. On the one hand, they would want their shared supporters to vote for them both, in order to maximize their respective totals and minimize the chance that either (or both) would lose to rivals from the other party. On the other hand, any candidate who aspired to the presidency would need some votes to be denied his running mate in order to ensure his own victory for the premier office.

This tension was felt before 1800 and by Federalists as well as Republicans. In 1788, when George Washington was the universal choice to become the first president, his lieutenant Alexander Hamilton worked secretly to disperse votes from John Adams, ostensibly in order to avoid embarrassing Washington with a close result.[10] Washington received votes from all 69 electors, while Adams got just 34, with the remaining 35 scattered among ten other notables. Adams was "deeply hurt" by his relatively modest support and later felt "sickened" when he learned that his total had been reduced by what he called a "dark and dirty intrigue." (McCullough 2001, 394, 409) In the 1792 election, which resulted in re-election of Washington and Adams, Hamilton did not need to divert votes from Adams. Although Washington continued to enjoy unanimous support (132 votes), the emergence of what soon became the Republican Party resulted in a significant challenge to the Vice President from George Clinton of New York, who received 50 votes to Adams' 77.[11]

In 1796, with Washington stepping down, a partisan contest ensued between Adams and Jefferson. With the vote expected to be close, each side sought regional balance in picking a vice presidential running mate–Thomas Pinckney of South Carolina for Adams, and Aaron Burr for Jefferson. The tension inherent in the approval voting system emerged with a vengeance. On the Federalist side, Hamilton was suspected of promoting the more malleable Pinckney over Adams for the presidency when he urged the strongest possible support for the South Carolinian –ostensibly to prevent Jefferson from finishing second (Mc-

---

[10] It may be that Hamilton also did not want Adams to acquire enough prestige to threaten his own ascendance within Washington's administration.

[11] In addition, Jefferson received four votes and Burr one–the latter (ironic in view of later events) a cut of one vote in South Carolina from an otherwise solid Washington-Adams tally.

Cullough 2001, 463; van der Linden 1962, 9-10). As Hamilton hoped, the eight electors from South Carolina crossed party lines to vote for both Jefferson and their favorite son, so Hamilton's plan would have succeeded had not eighteen Federalist electors in three New England states foiled it by cutting Pinckney, who finished with 59 votes, twelve behind Adams' 71. Jefferson was a close second with 68. Thus, in the last hurrah for original constitutional intent, the two most distinguished candidates were chosen president and vice president, even though they were partisan opponents. Meanwhile, on the Republican side, Burr suffered the unkindest cuts of all when 38 of Jefferson's 68 electors threw away their second votes on various alternatives rather than support the New Yorker. All but one of the 38 defectors were from Southern states, including nineteen of twenty electors from Jefferson's Virginia.[12]

As had Adams in 1788, Burr felt humiliated by the low vote caused by his betrayal at the hands of Southern Republicans. In 1800, Jefferson, driven by electoral necessity and showing considerable chutzpah, sought to revive the Republicans' Virginia-New York alliance by asking Burr to serve again as his running mate. The latter was willing to stand only "if assurances can be given that the southern states will act fairly." Jefferson's emissary, Albert Gallatin, pledged that they would (Wills 2003, 71). Soon afterwards, a meeting of Republicans in Philadelphia duly endorsed a ticket of Jefferson and Burr. Any breach of the bargain would tarnish the personal honor of leaders on both sides, given the strong commitments they had made (Freeman 2002). Nevertheless, both wings of the Republican party were in an agony of anxiety that the other would renege, and each frequently reminded the other of its pledge. James Madison personally made sure that none of his fellow Virginia electors cut Burr. As he noted, confidence (conveyed by Burr's emissaries) that New York would hold for Jefferson impelled Virginia "to give an unanimous tho' reluctant vote for B. as well as J." To ensure that no Burr ally in New York would drop a vote from Jefferson, one of the latter's loyalists there, General Philip van Cortlandt, "sportively insisted" on preparing each of his fellow electors' ballots for their signatures (van der Linden 1962, 229-32, 260).

With the two biggest Republican states thus voting in lockstep, could not a vote or two have been dropped from Burr in a smaller state? Jefferson expected that to happen. After a trip through New England, Burr sent assurances that Jefferson would win one or two votes more than himself in Rhode Island or Vermont (Daniels 1970, 218; van der Linden 1962, 179). Whether Burr's prediction was an honest mistake or a deliberate deception, both states ended up in the Federalist camp. Jefferson may also have anticipated a reprise of 1796 in South Carolina. To appeal to that pivotal state, the Federalists had

---

[12] Why did Southerners cut Burr? Although aspersions against his character already abounded in 1796, he was also "a lifelong opponent of slavery" (Wills 2003, 77).

nominated another favorite son, Charles Cotesworth Pinckney, a cousin of their 1796 candidate. Hamilton was again intriguing for a Jefferson-Pinckney vote there, but he overplayed his hand by writing a vicious attack on his own party's President, John Adams.[13] Burr obtained a copy of Hamilton's pamphlet, which was intended for limited distribution, and released it to the press. The resulting backlash stiffened Federalist support for Adams. It may also have motivated C. C. Pinckney's principled decision to spurn a deal with his state's Republican legislators, insisting instead on equal electoral votes for himself and Adams (van der Linden 1962, 239-43).

Thus, aside from constitutional and communications obstacles to coordination, dropping votes to prevent a tie risked two grave dangers for Republicans: (a) In light of the three-vote margin in 1796, uncertain prospects in several states right up until December 1800, and Hamilton's machinations on behalf of Pinckney, throwing away votes from either member of the ticket could easily have resulted in victory for one or both of their Federalist opponents. For example, if Republicans in South Carolina had repeated the 1796 cross-party vote, Pinckney would have defeated Burr for the vice presidency. A split there combined with one vote cut from Jefferson in any other state would have created a tie for the presidency between Pinckney and Jefferson. The House might well have resolved such a deadlock in favor of Pinckney (Ackerman and Fontana 2004). (b) Even more perilous, as Joanne Freeman (2002, 105-6) emphasizes, was the risk–given the intense mistrust between Jefferson and Burr–that to "drop a vote would be to invite retributive vote dropping elsewhere, thereby destroying whatever national party unity existed, and probably throwing the election to the Federalists." As the Federalist Uriah Tracy gloated, "It is really pleasant to see the Democrats in such a rage for having acted with good faith....Each declaring if they had not had full confidence in the treachery of the other, they would have been treacherous themselves; and not acted as they promised to act at [Philadelphia] last winter, (viz.) all vote for Jefferson & Burr." (van der Linden 1962, 250)

3.   Generalizing from the Election of 1800

Does the unhappy experience of 1800 portend danger if approval voting in its modern form were widely adopted for hotly contested elections? To address that question, this section will (a) extract the Burr Dilemma as a general version of the strategic problem in 1800, (b) argue that differences between the early U.S. voting system and modern approval voting do not negate the contemporary relevance of the Burr Dilemma, and (c) in light of the dilemma,

---

[13] Adams had alienated Hamilton and other "High Federalists" by successfully resisting war with France and dismissing Hamilton's closest allies from the Cabinet.

analyze the strategic situation of candidates under approval voting as a game of Chicken.

3.1   The Burr Dilemma

To apply the problem revealed in 1800 to approval voting generally, it will be helpful to state a generalized version of the Burr Dilemma:[14]

> *When three or more candidates compete for an office that only one can win, and voters (V) may support two (or more) of them by casting equal (approval) votes, candidates (C_1 and C_2) seeking support from the same group (G) of voters will maximize their respective votes if all members of G vote for both C_1 and C_2. Both candidates thus have an incentive to appeal for shared support. However, if such appeals succeed completely and neither candidate receives votes from members of $V - G$, the outcome will be at best a tie in which neither C_1 nor C_2 is assured of victory. Each candidate therefore has an incentive to encourage some members of G to vote only for himself or herself. If both C_1 and C_2 successfully follow such a strategy, either or both may receive fewer votes than some other candidate C_3 supported by members of $V - G$. The risk that both C_1 and C_2 will lose is exacerbated if a retaliatory spiral increases the number of single votes cast by members of G. At the limit, such retribution reduces approval voting to conventional single-vote balloting among the members of G or, if the problem is endemic, among all voters. The nearer that limit is approached, the lower the probability that advantages claimed for approval voting will be realized.*

3.2   Differences Between Approval Voting and the Method of 1788-1800

As I noted earlier, the system of presidential elections originally established by the Constitution differs in four ways from modern approval voting: (a) Presidential electors were required to vote for exactly two candidates, whereas approval voting allows voters to support as many candidates as they choose, from one to everyone running. (b) Approval voting is usually recommended for single-winner elections, whereas the electors' vote could determine two winners, the president and vice president. (c) Approval voting is typically combined with the plurality decision rule, whereas the Constitution requires an absolute majority. (d) The Electoral College in 1800 consisted of only 138 voters, whereas advocates of approval voting are most concerned to see it adopted in mass popular elections.

---

[14]The dilemma also applies when approval voting is used to decide policies rather than offices, and to situations where more than two candidates or options appeal to the same group.

Do any of these features of the system of 1788-1800 eliminate or substantially undermine the relevance of the Burr Dilemma to modern approval voting? I contend that they do not. Let us consider each in turn.

Exactly two votes. The ability to vote for *more than two* candidates might change electoral dynamics by encouraging entry of additional candidates seeking the support of group G, by giving candidates based on G hope that they might find votes in $V - G$, and by encouraging candidates based on $V - G$ to seek votes in G. By complicating the situation and making prediction more difficult, those possibilities reduce the perceived likelihood of a tie. They do not, however, affect the basic tension summarized in the Burr Dilemma–both C_1 and C_2 are motivated to ask members of G who favor the other candidate to double vote while encouraging their own supporters to bullet vote. As for the possibility of voting for *only one* candidate, electors in 1788-1800 could–and did–perform the functional equivalent by "throwing away" their second votes on a dignitary who was not seriously competing.

Two winners. Anticipating the possibility of throwing away votes, the Framers invented the office of vice president in part to reduce that temptation, by making the second vote count for something. After parties developed, most electors wanted to keep the vice presidency within their own party, so the incentive to vote a straight ticket was high, thus increasing the possibility of a tie. Nevertheless, everyone saw the presidency as the real prize–including the first two vice presidents. "The most insignificant office that ever the invention of man contrived" said Adams of his role as Washington's understudy (McCullough 2001, 447). Similarly, on taking second place to Adams in 1796, Jefferson commented, "[It] is the only office in the world about which I am unable to decide in my own mind whether I had rather have it or not. A more tranquil & unoffending one could not have been found for me. It will give me philosophical evenings in winter & rural days in summer." (van der Linden 1962, 32-3) Consequently, temptation to win the greater office by exploiting weaknesses of the voting system affected even those candidates who agreed to compete just for the vice presidency–whether due to a candidate's opportunistic ambition, for which Burr was fairly or unfairly blamed, or to manipulation by others, as with Hamilton and the two Pinckneys, and the Federalists and Burr. Thus the existence of a consolation prize does not change the basic character of elections in 1788-1800 as contests dominated by the choice of a single winner.

Majority rule. Early in the development of approval voting, Brams and Fishburn (1983, 42) warned that "coupling approval voting to a runoff system [which requires an absolute majority] ... produces a combination that is never strategyproof." Merrill and Nagel (1987) extended that warning, showing the strategic pitfalls of combining approval balloting with any decision rule other than plurality, and recommending (pending further study) only the "approval plurality" combination. The Constitutional requirement of an absolute majority makes the Electoral College a runoff system in which the contingent second stage is a vote by a quite different electorate (the House), conducted under bizarre rules. The majority requirement did not cause the impasse of 1800, however, because both Jefferson and Burr surpassed fifty percent of the vote. Instead, it created a second way the Electoral College could have been indecisive, but the possibility of a tie would have existed even if the decision rule had been plurality. The contingent election in the House also made a tie worse for the Republicans than other conceivable tie-breakers, such as flipping a coin, because it put them at the mercy of their enemies, the Federalists. Although other tie-breakers might be less dangerous, they would not change the motivational source of the Burr Dilemma–any candidate who aspires to win prefers an outright victory to a tie.

Small electorate. Obviously, the probability of a tie is greater in an electorate of 138 members than in one with 100,000,000 voters. Is the Burr Dilemma therefore realistically a problem only for small electorates? I think not, for two reasons.

First, although the Burr Dilemma is revealed most dramatically when the strategy of approving two candidates produces a tie between them, ambitious candidates choosing strategies *ex ante* will want not only to avoid a tie, but also its more general equivalent, a .5 probability of victory. In a large electorate, the probability of a literal tie may be exceedingly small, but if two candidates competing for the same base both scrupulously encourage double voting, it may be that neither will have an advantage–each will have a 50-50 chance of winning. To gain an edge–a higher probability of finishing first–each therefore has an incentive to encourage at least some supporters to bullet vote.

Second, the tie in 1800 was due less to the small size of the Electoral College than to the cohesion of its Republican members (even if that cohesion was the product of mutual mistrust). The 73 Republican electors responded to directions given, ultimately, by just two leaders–who in turn were vigilantly watching each other's every move. Heretofore, analysts have usually approached the question of strategic choices under approval voting from the viewpoint of individual voters deciding atomistically within a relatively large electorate (e.g., Brams and Fishburn 1983, Merrill 1988). If instead most voters respond to the cues of a few leaders, then even a large electorate may be analyzed as a game affected by strategic moves of those players (plus, in most cases, a random element).

The tendency of ordinary voters to follow directions from leaders, parties, or other groups increases with the complexity of voting choices. For example, in Australian Senate elections, the great majority of voters (almost 95% in 1998) mark their ballots to follow party recommendations, because use of the preferential ballot, combined with the requirement of a complete ordering of candidates, can compel voters to rank literally scores of candidates (Sharman, Sayers, and Miragliotta 2002, p. 552).[15] Similarly, in American cities using the "long ballot" (electing many officeholders, major and minor, at a single election), the power of party leaders derives from voters' willingness to use the "big party lever" or "sample ballot" endorsement cards, because they otherwise have no idea whom to support for most offices. In an approval-voting election with k candidates, each voter has $2^k - 1$ possible strategies (excluding the choice of approving no one). (Brams and Fishburn 1983, 27) A field of eight candidates–not uncommon in early U.S. presidential primaries–entails 255 voting options! Even in the simplest case, a three-candidate race, the voter must decide among seven approval-voting options. Consequently, many voters will respond to leaders' cues not just out of loyalty and admiration, but also to reduce the cognitive burdens of voting. In the next section, I therefore analyze approval voting from a different angle, as a game of strategy played by leaders.

3.3   Strategic Analysis of Approval Voting

In its simplest version, the Burr Dilemma in approval voting can be analyzed as a two-person game in normal form. Assume a three-candidate race to be decided by a set V, consisting of v voters. (Throughout, upper-case letters will designate individuals and sets, while lower-case letters will indicate the number of members in a set.) Candidates C_1 and C_2 appeal to voters belonging to group G, $G \subset V$, while C_3 is supported by members of $\{V - G\}$. Members of $G_1 \subset G$, prefer C_1 to C_2 while those in another subset, G_2, prefer C_2.[16] All members of G strongly favor both C_1 and C_2 over C_3. For the present, we assume that neither C_1 nor C_2 is acceptable to any member of $\{V - G\}$, all of whom will cast bullet votes for C_3. The candidates believe that $g > v - g$, but also believe that margin is small enough that $g - g_1 < v - g$, and $g - g_2 < v - g$. That is to say, neither C_1 nor C_2 can beat C_3 without votes from the other's supporters. C_1 and C_2 each have two strategies. They can ask their supporters to cast two votes, one for each candidate, or they can recommend a bullet vote. If all voters follow their leaders' cues, the possible outcomes are as shown in Table 1.

C_1's preference order over the four outcomes is $y > w > x > z$. C_2 ranks them $x > w > y > z$. This configuration defines the game known as Chicken, after a scenario in which two teenage thrill-seekers drive their cars straight at each other.[17] The first to swerve is derided as a weak and cowardly "chicken," whereas the one who continues straight on wins admiration

---

[15]The Australian Senate is elected using the single-transferable vote.
[16]Besides G_1 and G_2, there may be other members of G who are strictly indifferent between C_1 and C_2. Approval voting is ideal for them.
[17]For details about Chicken on which my account draws, see Dixit and Skeath 1999. See also Brams 1994.

Table 1. The Burr Dilemma as a Game of Chicken

C_2 Recommends C_1 Recommends	Vote for C_1 and C_2	Vote for C_2 Only
Vote for C_1 and C_2	w C_1 and C_2 tie (or, more generally, each has a .5 chance of winning)	x C_2 wins
Vote for C_1 Only	y C_1 wins	z C_3 wins

for reckless courage. Each driver most prefers the outcome where he continues straight while the other swerves (y for C_1, x for C_2), but if both continue straight, they create a mutual disaster–death or injury in a head-on collision (z). In between, each prefers being thought no worse a chicken than the other (w, mutual swerving) to the humiliation of being the only chicken while the other gets bragging rights (x for C_1, y for C_2). In approval voting, the counterpart of driving straight is recommending a vote only for oneself, while the equivalent of swerving is the cooperative strategy of encouraging one's followers to cast a second vote for another candidate.

Unlike the more famous Prisoner's Dilemma, there is no dominant strategy in Chicken. Each player's best choice varies depending on what the other is expected to do. If C_1 knows that C_2 will recommend a double vote, she can win by asking her followers to vote only for herself. Conversely, if she knows that C_2 will not share votes, she can deny the other party (C_3) a victory, but only by handing the prize to her in-group rival, C_2. Mutatis mutandis, the same holds for candidate C_2. As an instance of Chicken, the Burr Dilemma thus has two Nash equilibria, outcomes x and y, each to the advantage of a different candidate. In determining which (if either equilibrium) will occur, there is a strong first-mover advantage. Whichever candidate can credibly commit, before the other, to a bullet-vote strategy forces the second mover to relinquish victory in order to prevent the triumph of the third candidate whom they both oppose.

However, when the actors are emotional human beings, as opposed to abstract rational choosers, there is no guarantee that either Nash outcome will occur. If C_1 commits to a single-vote strategy, C_2 (or his followers) may refuse to meekly cede victory to an uncooperative rival who has put personal ambition ahead of group welfare. Following C_2's lead or acting on their own resentment, members of G_2 may also bullet-vote, like their counterparts in G_1. If so, approval voting collapses into conventional single-vote balloting; the majority is split between two candidates; and a candidate desired by only a minority wins. Conversely, as happened in 1800, both G_1 and G_2 may adhere to the two-vote strategy as each reminds the other of "the benefits of reciprocity and reputation." (Dixit and Skeath 1999, 213) The result of such mutual "swerving" will be a tie in which the group wins, but the individual candidates must take their chances with whatever mechanism exists for breaking a deadlock.

Whereas resentment can undermine the two Nash equilibria, the other two outcomes (w and z) are unstable precisely because they are not equilibria. The temptation to defect can upset w (mutual cooperation) as either or both groups try to grab an opportunistic win by bullet voting. Conversely, z (mutual defection) can give way to the temptation to capitulate by casting double votes in order to stave off victory by the third force. Any outcome can happen, and all are precarious.[18]

It may be objected that this pessimistic conclusion depends on two excessively simplistic assumptions–dichotomous strategies and uniform behavior within groups. To the contrary, both assumptions can be relaxed without eliminating the Burr Dilemma.

Dichotomizing leaders' strategies assumes that they have only two choices–recommend that all followers cast two votes, or recommend that all cast just one vote. In a real election the more likely strategy is the one both sides feared in 1800–leaders publicly affirm their commitment to two votes while surreptitiously asking a few supporters to cut the rival candidate. Such duplicity would be difficult to keep entirely secret or unsuspected, especially in a large group. If C_2 learned that C_1 was trying to prevent a tie by having just one of her supporters cut him, then he should respond by having two of his own backers cut C_1. She could retaliate by arranging three cuts. Eventually (perhaps quite soon, as in 1800), the retaliatory spiral would result in C_3's defeating both C_1 and C_2. Thus the second choice for each candidate can be read as "Some Supporters Vote for C_i Only" without changing the essential nature of the problem.

The tie outcome (x) results because everyone in G casts a double vote (unless prompted otherwise by their leaders) while no one outside of G votes for either C_1 or C_2. Such sharp divisions are plausible only in a relatively small, highly disciplined voting body–such as the Electoral College in 1800 (and also many legislatures). Relaxing the assumption of uniform behavior, however,

---

[18]This conclusion is compatible with, though independent of, Saari and van Newenhizen's (1988a, b) more general criticism that virtually any outcome is possible under approval voting with the same preference profiles, depending on the approval strategies voters choose to employ. For replies, see Brams, Fishburn, and Merrill 1988 and Brams and Sanver 2006.

does not change the basic strategic problem, as long as the cuts in G and gift votes from $\{V - G\}$ are unpredictable or predicted to affect both C_1 and C_2 equally (within the margin of polling error). Such cuts and gifts act as a random vote generator, making an actual tie less likely; but the expectation for both candidates remains an unpredictably close race, which they can most easily influence in their own favor by encouraging followers to bullet vote.[19]

The Burr Dilemma vanishes only if the candidates and voters know that exogenous cuts and gifts will favor one competitor significantly more than the other. Let us consider that possibility, while also relaxing the assumption that members of G_1 and G_2 always follow their candidate's cues. According to the Poll Assumption of Brams and Fishburn (1983, 115), given reliable poll information, utility-maximizing voters will vote for the more preferred of the two front-runners plus any trailing candidate(s) whom they prefer to that front-runner. In our three-person example, there are three possibilities, which I will consider from the viewpoint of G_1, voters who like C_1 best: (i) C_1 and C_2 are the front-runners. If their lead over C_3 seems secure, members of G_1 will vote only for C_1. Why cast a second vote that might help C_2 defeat their favorite if there is no danger that C_3 will win? (ii) C_1 and C_3 are the leaders. Again, a member of G_1 has no reason to help C_2. The only vote that has a chance to affect the outcome is the one he would cast for C_1 anyway. (iii) C_2 and C_3 top the poll, with C_1 trailing. Only in this case will a voter in G_1 cast two votes–a practical vote for C_2 to help him defeat the least-liked C_3 , and a sentimental or send-a-message vote for C_1.

If reliable polls thus identify a clear leader between C_1 and C_2, that candidate (say, C_1) can stay above the fray when it comes to offering advice about voting strategies. She does not want her followers to give a vote to C_2, because doing so increases the risk that she will end up in a tie with him; but she does not need to promote a bullet-voting strategy, because her supporters will reach that conclusion for themselves, if they behave according to the Poll Assumption (i.e., they learn about poll results and calculate rationally). C_2, on the other hand, is in a bind. If he pushes for bullet voting, many of his followers may nevertheless double vote or even desert him entirely for fear of wasting their votes. He spares them that dilemma and thus maximizes his own vote by

---

[19] Samuel Merrill (2004) points out that if candidates care only about their own individual prospects, putting no value on victory for their group, then their ranking of outcomes follows the pattern of a Prisoner's Dilemma, rather than Chicken. C_1, for example, would prefer z to x, because she reasons that bullet-voting by her own supporters maximizes her chance of defeating the in-group rival, C_2, without reducing her vote relative to C_3. By similar reasoning, C_2 prefers z to y. I grant that such purely egocentric motivation may sometimes occur, but when it does, approval voting faces an even greater problem than under the Chicken pattern. If candidates conform to Prisoner's Dilemma motives, encouraging bullet-voting is the dominant strategy for each candidate, z is the sole Nash equilibrium, and an approval voting election degenerates into single-vote plurality. A real election in which such degeneration occurred is described in Saari 2001a.

endorsing a two-vote strategy; but as long as C_1 has no need to reciprocate, C_2 is condemned to finish as an also-ran.

The preceding scenario is what advocates of approval voting usually have in mind. Ability to support more than one candidate enables voters "to give the devil his due" (e.g., Ralph Nader in 2000 or 2004) while preventing him from acting as a spoiler. Would the benefits approval voting offers in such situations offset the problems it might cause when the Burr Dilemma occurs? To make such a judgment, it is necessary to consider not only the relative likelihood of each scenario, but also whether, in view of the Burr Dilemma, a different electoral reform might be more promising than approval voting.

4.   Conclusion

The mere fact that approval voting runs into a problem with strategic behavior is not sufficient to reject it. We know from the Gibbard-Sattherthwaite theorem that, when there are three or more choices, *all* voting procedures are vulnerable–not always, but under some configurations of preferences–to manipulation by strategic voting (Gibbard 1973). Therefore, simply showing the possibility of such a problem should never suffice to discredit a voting system. Choosing among voting methods requires comparative judgments involving numerous criteria, of which discouragement of (or vulnerability to) strategic voting is only one. Others include simplicity, economy of administration, likelihood of majority rule (Condorcet efficiency), avoidance of highly unpopular outcomes (Condorcet losers), resistance to spoilers (independence of irrelevant alternatives), and positive responsiveness to voters' choices (monotonicity). Moreover, in evaluating manipulability through strategic voting, analysts must go beyond its mere possibility to assess various forms of strategic voting and their relative undesirability, obviousness, ease, and probability under alternative voting systems. Thus recognizing that the Burr Dilemma can occur does not necessarily entail rejecting approval voting, especially as the method can function well when the conditions required for the Poll Assumption obtain. Nevertheless, the problem is sufficiently serious as to require more careful analysis of approval voting in comparison with other options for reform of single-winner elections.

References

Ackerman, Bruce, and David Fontana. 2004. "How Jefferson Counted Himself In." *The Atlantic Monthly* 293:2 (March): 84-95.

Brams, Steven J. 1994. *Theory of Moves*. Cambridge, UK: Cambridge University Press.

Brams, Steven J., and Peter Fishburn. 1983. *Approval Voting*. Boston: Birkhauser.

Brams, Steven J., Peter Fishburn, and Samuel Merrill III. 1988. "The Responsiveness of Approval Voting: Comments on Saari and van Newenhizen." *Public Choice* 59: 112-31.

Brams, Steven J., and M. Remzi Sanver. 2006. "Critical Strategies under Approval Voting: Who Gets Ruled in and Ruled out." *Electoral Studies* (forthcoming)

Ceaser, James W. 1979. *Presidential Selection: Theory and Development*. Princeton: Princeton University Press.

Daniels, Jonathan. 1970. *Ordeal of Ambition: Jefferson, Hamilton, Burr*. Garden City, NY: Doubleday.

Dixit, Avinash, and Susan Skeath. 1999. *Games of Strategy*. New York: W.W. Norton.

Freeman, Joanne B. 2002. "Corruption and Compromise in the Election of 1800: The Process of Politics on the National Stage." In James Horn, Jan Ellen Lewis, and Peter S. Onuf, eds., *The Revolution of 1800: Democracy, Race, and the New Republic*. Charlottesville: University of Virginia Press, 2002, 87-120.

Gibbard, Allan. 1973. "Manipulation of Voting Systems: A General Result." *Econometrica* 41: 587-601.

McCullough, David. 2001. *John Adams*. New York: Simon and Schuster.

Merrill, Samuel III. 1988. *Making Multicandidate Elections More Democratic*. Princeton: Princeton University Press.

Merrill, Samuel III. 2004. E-mail to the author, May 21.

Merrill, Samuel III, and Jack H. Nagel. 1987. "The Effect of Approval Balloting on Strategic Voting under Alternative Decision Rules." *American Political Science Review*. 81:2 (June): 509-24.

Randall, Willard Sterne. 1993. *Thomas Jefferson: A Life*, New York: Henry Holt & Co.

Saari, Donald G. 2001a. "Analyzing a Nail-Biting Election." *Social Choice and Welfare*. 18: 415-30.

Saari, Donald G., and Jill van Newenhizen. 1988a. "The problem of indeterminacy in approval, multiple, and truncated voting systems." *Public Choice* 59: 101-20.

Saari, Donald G., and Jill van Newenhizen. 1988b. "Is approval voting an 'unmitigated evil'?" *Public Choice* 59: 133-47.

Sharman, Campbell, Anthony M. Sayers, and Narelle Miragliotta. 2002. "Trading Party Preferences: The Australian Experience of Preferential Voting." *Electoral Studies* 21:543-60.

Van der Linden, Frank. 1962. *The Turning Point: Jefferson's Battle for the Presidency*. Washington: Robert B. Luce, Inc.

Wills, Garry. 2003. *"Negro President": Jefferson and the Slave Power*. Boston: Houghton Mifflin.

The Italian Bug: A Flawed Procedure for Bi-Proportional Seat Allocation

Aline Pennisi

Abstract   There is a serious technical flaw in the newly approved Italian electoral law. The flaw lies in the method used to allocate the Chamber of Deputies seats to parties (or coalitions) within multi-member regional constituencies. The procedure stated in the law could produce contradictory results: it could end up assigning a party more (or less) seats than it is entitled to receive on the basis of the same law. At least two types of paradoxes may occur. Although they have been utterly overlooked in the debate over electoral reform, they can be critical in practice when trying to determine the actual seat allocation. The failure of the current Italian electoral law was inherited from the previous one but the consequences are worse. Moreover, a correction mechanism introduced into the law at the last-minute does not prevent it from producing contradictory results. The paradoxes that undermine the Italian electoral law are pointed out and a solution is proposed. A broad conclusion is that a more extensive use of mathematics in the design and evaluation of electoral systems would help identify flaws and deliver more transparent, logical and fairer electoral laws.

Keywords: Bi-proportional allocation, Italian electoral system, proportional systems.

1.   Introduction

On the 14th of December 2005 Italy endorsed a new law[1] for the election of representatives at the Chamber of Deputies and Senate. The new electoral law replaced a fairly recent hybrid system[2] with a proportional one, which includes a threshold for parties to be eligible to receive seats and a (potentially big) majority prize for the party (or coalition of parties) with the most votes. The debate over electoral reform questioned the opportunity of introducing

---

[1] The initial proposal for a reform of the electoral system was presented on 13th September 2005 by the Polo coalition in the lower house (Chamber of Deputies), approved with some modifications in October and ratified by the upper house (Senate) on the 14th December.

[2] In the mixed system (L. 4 agosto 1993), also know as "Legge Mattarella" or "Mattarellum", 75% of the seats were assigned on first-past-the-post rules and the remaining 25% on proportional basis. It was first introduced in 1993, at the time of a major turmoil in Italian political setting caused by Tangentopoli and the consequent collapse of the traditional parties.

such a radical change with general elections forthcoming in spring 2006 and mainly focused on its political effects in terms of coalition strategies and party fragmentation. The public and the media utterly overlooked a serious technical flaw that the new law inherits from the old one. The procedure adopted to transform votes into seats has a "bug" and one could end up with paradoxical results: the law might award a party more (or less) seats than those it is entitled to by the same law! This has considerable practical consequences on how to decide the actual seat allocation and casts doubt on the legitimacy of the electoral law itself. The flaw has to do with the fact that, for some voting outcomes, the procedure will get stuck when distributing the seats among parties/coalitions within the regional constituencies for the Chamber of Deputies. The purpose of this paper is not to discuss whether it was appropriate to change the Italian electoral system but to prove one of its shortcomings from a purely technical point of view. Whatever the opinion on the law and modifications it introduces, one would surely agree that it should guarantee a consistent and unique outcome in terms of seats and attempt, to the extent to which this is possible, to guarantee fairness (i.e., citizens' votes should have the same weight in determining the electoral outcome). The transformation of votes into seats is a mathematical problem and in order to satisfy basic requirements of logic, transparency and equity among citizens it should be consistently defined and correctly solved in all circumstances. Unfortunately, the system under exam fails to do so.

The new Italian electoral law for the election of representatives of the Chamber of Deputies (Ddl Camera 2620 13, 2005) allocates seats proportionally to the votes obtained by each party (and coalition of parties) at the national level and within multi-member regional constituencies. A majority prize is meant to ensure that the party or coalition with the greatest number of total votes wins a full majority of seats in the Chamber of Deputies (i.e., at least 340 seats) no matter how many votes the other parties receive[3] There is a single ballot and candidates are elected on the basis of regional "blocked" lists (citizens do not express their preference for a candidate but a vote for a party list). Moreover, a complex scheme of thresholds is adopted to select which parties and coalitions are eligible to compete in the seat allocation. The Italian Constitution sets the size of the Chamber of Deputies at 630 seats. There are 27 multi-member regional constituencies in total. The Constitution also establishes that the number of seats at stake in each regional constituency must be proportional to the number of its inhabitants, according to the latest population census. The only exception is the region of Valle d'Aosta which has a single-member dis-

---

[3] The seat bonus represents 54 per cent of the seats, no matter what the weight of the strongest party in terms of votes, and basically introduces a majority component which undermines the proportional principle attempted in the law.

trict. Finally, 12 seats are assigned to a constituency of Italian citizens resident abroad[4].

The electoral reform of 2005 is not the first attempt to modify the Italian mixed system which previously allocated 75% of the seats in single-member districts with first-past-the-post rules and the remaining 25% on a proportional basis. Several other proposals have been made in the last decade, but surprisingly enough they usually sought to abolish the proportional seats and introduce a fully first-past-the-post system. On the other hand, the new electoral law adopts a proportional logic, although mitigated by the majority prize.

2.   Where the Italian System Fails

The Italian electoral law wishes to achieve a double proportionality: at the national level and within the regional constituencies. But the procedure implemented to achieve this is flawed and, for some voting outcomes, it will end up by awarding a party more (or less) seats within the regional constituencies than those the same party is entitled to at the national level. A similar flaw was identified by Balinski and Ramirez in the 1996 Mexican electoral law (Balinski and Ramírez González, 1997). In short, the new electoral law first allocates seats to parties at the national level and then assigns seats to the parties within each regional constituency. Both steps are carried out on a proportional basis, according to a method called *Hare* or *Largest Remainders*. A fundamental property of the method is that the number of seats is always equal to the exact share (quota) of seats a party should receive on a proportional basis, either rounded down or rounded up. There is an extensive literature on proportional electoral systems and the Largest Remainders method, for details see for example (Balinski, 2004; Balinski and Young, 1982; Grilli di Cortona et al., 1999).

Since the computation of the number of seats to each party or coalition at the national level is carried out first, the allocation of seats to parties within the regional constituencies is bound to satisfy two sub-totals: (a) the sum of the seats assigned to all parties within a given constituency must be equal to the number of seats actually at stake in the constituency and (b) the sum of the seats awarded to a given party in all constituencies must be equal to the number of seats it was awarded on the basis of the national computation.

The procedure adopted to allocate seats to parties within the regional constituencies starts by computing the exact number of seats due to each party one constituency at a time (starting from the smallest one). This number is equal to the size of the constituency multiplied by the percentage of ballots the given party has obtained. This "exact quota of seats" is not necessarily an integer

---

[4]Voting by Italian nationals resident abroad is governed by L. 27 dicembre 2001 n. 459 (known as "Legge Tremaglia") and by its implementing regulation (D.P.R. no.104 / 2003).

and usually carries a fractional part. Since a seat cannot be divided among different candidates, the law first assigns each party a number of seats equal to the exact quota rounded down. If there still remain seats to be assigned, these are awarded to parties in the order of the largest fractional remainders.

The problem with this procedure is it does not guarantee that, once all seats are assigned, the total amount awarded to each party is the same as the amount computed at the national level. Basically, by operating one constituency at a time, without worrying about the total amount of seats a party is entitled to at the national level, the sum constraint might not be satisfied. This is not a negligible defect and it has serious practical consequences: should such a paradoxical result occur, who will decide the final seat assignment? The size of the Chamber of Deputies cannot be changed. Some parties will gain more seats with the regional allocation but others will with the national one. The failure of the Italian electoral law could trigger a serious controversy between political parties on whether the result of the national allocation should prevail on the results of the regional allocations. Claims of the different political groups would presumably vary according to which case is the most advantageous for them.

3.   Electoral Paradoxes

Small scale examples of the Italian electoral paradox have been already discussed in (Pennisi et al., 2005a) but more realistic examples can easily be produced. Consider the case of six political parties competing for the 617 seats at stake in the Italian Chamber of Deputies and the 26 regional constituencies[5]. Let the voting outcome be the one detailed in Table 1. In this example the number of votes is comparable with those expressed by Italian citizens in the last general elections held in 2001: there is, at the most, a 4.5 point difference between the share of party votes shown in the example and those obtained in 2001. Notice that party C is competing only in some constituencies. This was the case of the Northern League in the 2001 elections. Let the party with the greatest number of votes be the "majority list" and the quotient between the total number votes and the number of seats at stake (617) be called the fractional national coefficient. This number rounded downwards is called the national coefficient and represents the "cost" of a seat in terms of votes in the national contest.

---

[5]The actual size of the Chamber of Deputies is 630 seats, as established in the Italian Constitution (article 56), but 12 of them are reserved to the election of representatives of the Italians living abroad and 1 to the single/member district of Valle d'Aosta.

Table 1.   Number of votes per party and constituency.

Constituency	Party A	Party B	Party C	Party D	Party E	Party F	Total votes	Seats at stake
Piemonte 1	400783	96054	73072	249544	237700	30383	1087536	24
Piemonte 2	285589	124753	194327	180623	132317	33655	951264	22
Lombardia 1	761543	249386	160114	321368	396133	48568	1937112	40
Lombardia 2	636642	192461	410789	192385	357854	97506	1887637	43
Lombardia 3	186671	78789	127148	170008	79445	1607	643668	15
Trentino Alto Adige	82778	55049	0	51517	67413	36885	293642	10
Veneto 1	329782	158402	169965	177987	251340	8289	1095765	29
Veneto 2	317041	89550	143286	131320	180218	44223	905638	20
Friuli Venezia Giulia	186371	96356	67321	66763	146959	1847	565617	13
Liguria	280399	84842	77958	249392	123161	43594	859346	17
Emilia Romagna	641699	307914	0	839563	469029	266707	2524912	43
Toscana	508202	350951	0	774294	360079	219358	2212884	38
Umbria	134556	134860	0	183441	113560	83630	650047	9
Marche	244594	179000	0	252977	183337	95155	955063	16
Lazio 1	564330	559634	0	473675	491237	173156	2262032	40
Lazio 2	307780	200949	0	183038	126949	84146	902862	15
Abruzzi	239180	157899	0	179005	131944	84537	792565	14
Molise	71393	59449	0	71950	59301	48741	310834	3
Campania 1	567890	239670	0	312391	197204	129222	1446377	33
Campania 2	463265	250225	0	212514	261133	101904	1289041	29
Puglia	681996	397202	0	341107	415243	150335	1985883	44
Basilicata	103244	72139	0	100214	101297	56334	433228	6
Calabria	270118	195793	0	223719	150159	97080	936869	22
Sicilia 1	478296	150643	0	190233	188381	88548	1096101	26
Sicilia 2	504881	221958	0	171734	271583	82452	1252608	28
Sardegna	301315	153912	0	209921	166320	126415	957883	18
TOTAL	9550338	4857840	1423980	6510683	5659296	2234277	30236414	617

The procedure adopted by the law first allocates the 617 seats to parties at the national level taking into account the majority prize, as follows[6]:

- assign to each party its exact quota of seats rounded down, i.e. divide the number of votes the party has obtained by the national coefficient and round this number down. Then count the number of seats that must still be awarded and assign an additional seat to those parties which have the greatest fractional remainders (this is a slight variant of the typical statement of the largest remainders method);

- check whether the majority list has achieved at least 340 seats;

---

[6]The whole setting is slightly more complicated because of the national and regional thresholds on the number of votes parties must obtain to compete in the electoral contest. For the sake of simplicity, suppose all parties in the example satisfy the thresholds. The technical flaw put forth in this paper is not related to the use of thresholds, although in paragraph 4 we do suggest that thresholds may play a role in making the paradoxes more likely to occur.

- if this is not the case, assign 340 seats to the majority list and calculate the majority electoral coefficient (i.e., the total majority list votes divided by 340 and rounded downwards) and the minority electoral coefficient (i.e., the sum of votes obtained by the other parties divided by 277 and rounded downwards). Distribute the 277 remaining seats among the minority parties with the method of largest remainders described above, but using the minority electoral coefficient.

The majority and minority coefficients represent the cost in terms of votes of a seat for the majority list and for all other parties, respectively. In a truly proportional electoral system the cost of a seat is the same for all parties, but here, the adoption of the 340-seat majority prize can introduce large differences: seats may cost much less for the majority list and much more for all other parties. In the example the cost of a seat for "minority" parties is 2.7 times larger that the cost of a seat for the majority list (approximately 28 thousand ballots are needed for the majority list to get a seat against about 75 thousand for the other parties!). The number of seats assigned to each party is given in Table 2.

Table 2.   Seats awarded at the national level, taking into account the majority prize.

	Party A	Party B	Party C	Party D	Party E	Party F	Total seats
Seats awarded at the national level	340	65	19	87	76	30	617

At this point, seats must be allocated to the parties within the regional constituencies. The procedure consists of the following steps, for each constituency:

- divide the number of votes obtained by the majority list by the majority coefficient and for each other party, divide the number of votes by the minority coefficient; these indexes are the relative costs of a seat in the constituency.

- multiply the number of seats at stake in the constituency by each party's index and divide this product by the sum of all indexes to obtain the exact number of seats assigned to each party, and round this number down;

- assign the remaining seats at stake in the constituency on the basis of largest remainders.

Despite the convoluted formulation, this procedure is nothing more than a slight variant of the largest remainders method, applied at the constituency level and taking into account two different totals (for majority and minority seats). The result is given in Table 3. The law's internal contradiction is clear: parties A, C and D end up with more seats than those they are entitled to at the national level while E and F have less. The only correct result is the number of seats assigned to party B.

In the bill for electoral reform presented in September 2005 the procedure ended at this point, totally neglecting the fact that a paradox - such as the one shown in the example - could occur. During parliamentary discussion the legislators must have realized, at the last moment, that something could go wrong. In a version of the law approved by the Chamber of Deputies in October a correction procedure was introduced. The idea underlying the correction mechanism is to re-balance the seat distribution through transfers of seats between parties with a surplus (in the example A, C and D) to parties with a deficit (E and F). Unfortunately, this mechanism is once again flawed!

The correction mechanism is executed whenever the sum of seats awarded to parties in the regional constituencies is not equal to the corresponding national seat allocation. It is applied starting from the party with the largest seat surplus, in decreasing order. Seats are transferred from the party with a surplus in those constituencies in which the party has obtained an additional seat thanks to its remainders, selecting the smallest remainders (the underlying idea is that seats are taken away from the party in those cases in which it was less entitled to them respect to other constituencies). The seats are transferred to a party with a seat deficit in the same constituency provided that such party has not already benefited from an additional seat on a remainder's basis and according to the largest unused remainder (the idea is to award the seat to the party which is next most entitled to it).

Although it is meant to correct the damage done, the mechanism does not always work because it operates only on seats rounded up, i.e. assigned to a given party thanks to its relatively "large" remainder. In other words the correction mechanism assumes that a paradox may occur, but only because a party has benefited too much from its exact quotas being rounded up. In Table 3 bold figures highlight exact quotas rounded upwards, i.e. cases in which an additional seat was awarded to the party during the regional allocation procedure thanks to the largest remainders. Note that party C is in surplus of seats although it has received only exact quotas rounded down. A double-star identifies cases in which a surplus party received an additional seat with relatively small remainders (the smallest among all remainders it has used) and a star identifies cases in which a deficit party has the largest unused remainders. These are the parties and constituencies involved in the seat transfers.

One can already notice that, despite transfer operations, the inconsistency between the sum of seats allocated to parties within the regional constituencies and the national allocation will still hold: party C will keep two extra

Table 3. Seats awarded to parties within the constituencies on the basis of largest remainders.

Constituency	Party A	Party B	Party C	Party D	Party E	Party F	Row sum
Piemonte 1	**15**	1	1	**4****	3	0*	24
Piemonte 2	**12**	**2**	3	**3**	2	0	22
Lombardia 1	25	3	2	4	**5**	**1**	40
Lombardia 2	**25**	**3**	6	**3**	5	1	43
Lombardia 3	**8**	1	2	**3**	1	0	15
Trentino Alto Adige	5	1	0	1	**2**	**1**	10
Veneto 1	**16****	**3**	3	3	**4***	0	29
Veneto 2	12	1	2	**2**	2	**1**	20
Friuli Venezia Giulia	7	**2**	1	**1**	2	0	13
Liguria	**10****	1	1	3	**2**	0*	17
Emilia Romagna	20	**4**	0	10	**6**	3	43
Toscana	**17**	**4**	0	**10**	4	**3**	38
Umbria	**4**	1	0	**2**	1	**1**	9
Marche	**8**	2	0	**3**	2	1	16
Lazio 1	**19**	**7**	0	**6**	6	2	40
Lazio 2	**9**	2	0	**2**	1	**1**	15
Abruzzi	7	**2**	0	2	**2**	**1**	14
Molise	1	**1**	0	**1**	0	0	3
Campania 1	**21**	3	0	4	3	**2**	33
Campania 2	17	**4**	0	**3**	**4**	1	29
Puglia	25	**6**	0	**5**	**6**	2	44
Basilicata	**3**	**1**	0	**1**	1	0	6
Calabria	11	3	0	**4**	2	**2**	22
Sicilia 1	17	2	0	**3**	**3**	1	26
Sicilia 2	**18**	**3**	0	2	**4**	1	28
Sardegna	**10**	**2**	0	**3**	2	1	18
Column sum	**342**	**65**	**21**	**88**	**75**	**26**	**617**
Seats awarded at the national level	**340**	**65**	**19**	**87**	**76**	**30**	**617**
Surplus and/or deficit	**+2**	**0**	**+2**	**+1**	**-1**	**-4**	

seats and party F will lack them. There is no mention in the law on how to resolve such situations. In fact, the law states that when seat transfers within a same constituency are no longer possible (just as in the example), in order to eliminate any other surplus, seats may be given to parties with a deficit in a different constituency, with the largest unused remainders first. However, it never acknowledges the fact that a surplus of seats may be due to the simple assignment of exact quotas rounded down and not to seats awarded according to largest remainders. Other realistic examples can be produced, as shown in (Pennisi et al., 2005b). There are at least two types of paradoxes undermining the Italian electoral law and for which its correction mechanism is not sufficient to repair:

- the *surplus paradox* for parties with exact regional quotas all rounded downwards: when the sum of the seats assigned to a party (or coalition of parties) in the constituencies is greater than the number of seats it is entitled to at national level and all its regional seats are the result of exact quotas rounded downwards;

- the *deficit paradox* for parties with exact regional quotas all rounded upwards: when the sum of the seats assigned to a party (or coalition) in the constituencies is smaller than the number of seats it is entitled to at the national level and it has already benefited from extra seats thanks to largest remainders in all constituencies where it has obtained votes.

The first type of paradox is shown in the example, the second is symmetric. In the second case the correction mechanism will get stuck because the law never considers the possibility that a lack of seats can occur although a party's exact quota of seats has already been rounded upwards in all constituencies (and therefore the party is never eligible to receive additional seats).

Moreover, applying the correction mechanism can cause a third type of paradox: the *constituency paradox*. In fact, given that seat transfers between parties in different constituencies are allowed, the total number of seats awarded in each constituency can end up being different from the number of seats actually at stake in the same constituency![7]

As mentioned earlier, the "bug" in the new Italian electoral law was inherited from the previous one. The proportional seat allocation of the hybrid system adopted for elections in 1994, 1996 and 2001 - also known as the Mattarellum system from the name of its maker - was in fact carried out with the same largest remainders procedure, applied first at the national level and then in the regional constituencies one at a time. There was no correction mechanism but while allocating additional seats to parties according to the Largest Remainders at the regional level, the number of seats each party was entitled to at the

---

[7]Actually, this occurred in the 2006 elections where 11 seats were assigned in Trentino Alto Adige - one more than the number of seats at stake in that constituency - and 2 seats were awarded in Molise instead of 3. Such a result is in clear contradiction with the Italian Constitution.

national level was considered an upper bound. Hence, the idea was to prevent a party from violating the constraint on the total number of seats it could receive by rounding its exact quota up only until it had not reached the number of seats already awarded at the national level. This does not prevent the paradoxes from occurring because, as we have stressed, even parties with exact quotas all rounded down can end up with a surplus and parties with exact quotas all rounded up can end up with a deficit.

In the new electoral law the technical flaw is even worse than in the old one, for several reasons. First of all, because it concerns the allocation of all 617 seats, harming the whole electoral outcome, while the Mattarellum system allocated only 25% of the seats on proportional basis. Secondly, because, in the case of coalitions competing in the electoral contest, the paradox occurring in trying to justify national and regional results may also occur in trying to re-allocate the seats awarded to the coalition among its member parties. Finally, the introduction of a correction mechanism while the bill was under exam in Parliament suggests that legislators saw a flaw in the procedure; the persisting failure of the correction mechanism proves they have not understood the real nature of the problem.

4.   Tackling the Italian Electoral Problem

From a mathematical point of view, put aside the majority prize the electoral procedure adopted in the Italian case is meant to solve the following problem: find a matrix of nonnegative integers (the seats), whose row (the constituencies) and column (the political parties) sums are fixed and whose entries are "proportional" to a given matrix (the matrix of votes). This is the well-known bi-proportional allocation problem in integers which is in itself of great interest and has many applications, not only in the electoral field (for example Bacharach, 1970; Balinski, 1989a; Leti, 1970). Let M be a set of regional constituencies, N a set of political parties (or coalitions) and s a positive integer equal to the total amount of seats to be allocated (or house-size). The following notation is used:

v_{ij} the number of votes for party j in constituency i;
s_i the number of seats at stake in constituency i and such that $\sum_{i \in M} s_i = s$;
t_j the number of seats awarded to party j at the national level;
v the total number of votes.

Then v_{iN} and v_{Mj} are respectively the sum of the votes cast in constituency i (across all parties) and the sum of the votes cast for party j (across all constituencies):

$$v_{iN} = \sum_{j \in N} v_{ij}$$

$$v_{Mj} = \sum_{i \in M} v_{ij}$$

$$v_{MN} = \sum_{i \in M j \in N} v_{ij} = V$$

The bi-proportional allocation problem in integers is to find a matrix of seats s_{ij} for each constituency $i \in M$ and each party $j \in N$ such that the following constraints hold:

$$
\begin{aligned}
s_{MN} &= s \\
s_{iN} &= s_i && \text{for every constituency } i \\
s_{Mj} &= t_j && \text{for every party } j \\
s_{ij} &\geq 0 && \text{for all } i, j \\
s_{ij} &\text{ integer} && \text{for all } i, j
\end{aligned}
\tag{1}
$$

Finally, one would like s_{ij} to be *"as proportional as possible"* to v_{ij} for all $i \in M$ and $j \in N$.

Let $q_{ij} = v_{ij} \frac{s_i}{v_{iN}}$ be the exact quota of seats for party j in constituency i. Now $q_{iN} = s_i$ and $q_{MN} = s$. Perfect proportionality is achieved by letting $s_{ij} = q_{ij}$. If there are no further constraints, this is the obvious solution to the problem, but s_{ij} must be integer and $s_{Mj} = t_j$ must hold as well. The idea underlying the Italian method is to consider the exact quotas each party is entitled to in the regional constituency and to round these numbers up or down (in the case the majority prize is assigned to some party, these quotas are not the exact ones but a modified version based on the majority or minority seats). Unfortunately, it is fairly easy to build realistic examples for which, however the rounding is carried out, it is impossible to satisfy both row and column constraints ($s_{iN} = s_i$ and $s_{Mj} = t_j$). The Italian electoral law adopts the method of largest remainders both at the national and regional level. Therefore, all resulting seat allocations comply with a property called quota satisfaction: i.e,

$$\left\lfloor v_{ij} \frac{s_i}{v_{iN}} \right\rfloor \leq s_{ij} \leq \left\lceil v_{ij} \frac{s_i}{v_{iN}} \right\rceil$$

holds for every party j and constituency i (at the regional level) but also:

$$\left\lfloor v_{Mj} \frac{s}{v} \right\rfloor \leq t_j \leq \left\lceil v_{Mj} \frac{s}{v} \right\rceil$$

holds for every party j at the national level.

The *surplus paradox* occurs if there is a party j such that:

$$\sum_{i \in M} \left\lfloor v_{ij} \frac{s_i}{v_{iN}} \right\rfloor > \left\lceil v_{Mj} \frac{s}{v} \right\rceil \tag{2}$$

The *deficit paradox* occurs if there is a party j such that:

$$\sum_{i \in M} \left\lceil v_{ij} \frac{s_i}{v_{iN}} \right\rceil < \left\lfloor v_{Mj} \frac{s}{v} \right\rfloor \tag{3}$$

For the paradoxes to occur there must be some kind of imbalance between vote/seat ratios at the national and regional level (or between the cost of a seat at the national and regional level), as shown below.

PROPOSITION 1 *If $s_i/v_{iN} = s/v$, for every $i \in M$ the two paradoxes cannot occur.*

Proof. If $s_i/v_{iN} = s/v$, for every $i \in M$, then for every party j:

$$\sum_{i \in M} \left\lfloor v_{ij} \frac{s_i}{v_{iN}} \right\rfloor = \left\lfloor v_{1j} \frac{s}{v} \right\rfloor + \left\lfloor v_{2j} \frac{s}{v} \right\rfloor + \ldots + \left\lfloor v_{Mj} \frac{s}{v} \right\rfloor \leq$$

$$\leq v_{1j} \frac{s}{v} + v_{2j} \frac{s}{v} + \ldots + v_{Mj} \frac{s}{v} = \frac{s}{v} \sum_{i \in M} v_{ij} \leq \left\lceil v_{Mj} \frac{s}{v} \right\rceil$$

which is the opposite of (2). The same can be shown for (3). ■

In other words discrepancy between national and regional coefficients is a necessary condition for the paradoxes, but it is not sufficient. Nevertheless, to get some intuition one may notice that when the regional seat apportionment plan is "perfect" - in the sense that the number of seats at stake in each constituency is perfectly proportional the corresponding regional population (fractional seats being allowed)- the "anti-paradox" condition $s_i/v_{iN} = s/v$ is equivalent to assuming the same rate of vote participation across the country.

PROPOSITION 2 *Let p be the total country population and p_i the population of the i-th constituency. Given a perfectly proportional seat apportionment plan, the condition $s_i/v_{iN} = s/v$ for every $i \in M$ is equivalent to the condition $v_{iN}/p_i = v/p$ for every $i \in M$.*

Proof. In a perfectly proportional seat apportionment plan $s_i/s = p_i/p$. If $s_i/v_{iN} = s/v$ holds for every $i \in M$, then $v_{iN} = s_i v/s = v p_i/p$ for every

$i \in M$ and vice-versa. ■

In real life a certain degree of discrepancy between the vote/seat ratios at the national and regional level is usual. Significant differences may be due to factors which are out of the legislator's control, such as different rates of absenteeism (citizens not voting at all) or protest (citizens casting invalid ballots) across the regions. As suggested in Proposition 2, they might also be due to the way the electoral law is put into practice, such as a "bad" regional apportionment plan (where the seats at stake in each regional constituency do not tend to reflect of the size of the constituency's population). Despite a "good" regional apportionment plan, there are at least two other features of the Italian electoral law that could be responsible for an imbalance between the national and regional ratios:

- The thresholds on the number of votes parties and coalitions must obtain to participate in the electoral contest. When a small party is cut out from the competition because of the threshold, its votes are deducted from the total constituency outcome in terms of votes. Parties running such a risk are typically groups of local interest which run only in very specific regions (at least in the Italian case);

- The majority prize. When the majority list wins 340 seats although it has obtained proportionally a much smaller amount of votes, the majority and minority coefficients tend to be very different, and different from the national vote/seat ratio.

Although the problem the Italian electoral law attempts to solve is not an easy one, a "sound" solution always exists as proved by Balinski and Demange (1989a and 1989b). The authors actually prove that a solution satisfying a number of basic properties (such as monotonicity, uniformity, relevance, exactness, etc.) can be found with an algorithm resembling the well-known out-of-kilter algorithm for minimum cost network flows.

5.   Drawing Some Conclusions

The history of electoral systems is full of examples of paradoxes and failures - some of which have been used with bias for the purpose of political advantage. A mathematical approach to electoral systems can help identify such failures. In fact a more thorough use of mathematical tools to evaluate and design the many features that make an electoral system - from the design of electoral districts to the choice of a method to transform votes into seats - is fundamental (see also Balinski and Young, 1982; Grilli di Cortona et al., 2005). The paradoxes underlying the Italian law are not due to the fact that achieving double proportionality, at the national level and within regional constituencies, is an unsolvable problem but to fact that the method adopted is not an appropriate one. In the case examined in this paper the legislators do not seem to have been aware of the underlying complexity of the problem they were facing and they have basically established a procedure which is too simple to address bi-proportional allocation in integers. Appropriate and correct procedures exist although they use somewhat sophisticated mathematics and might have to be carried out with the help of a computer program. This should not prevent electoral laws to adopt correct procedures: the University of Augsburg developed a Java-program for matrix apportionments using divisor methods and based on alternate scaling called BAZI, which has been adopted to shape the Zurich electoral law in 2003 (Pukelsheim, 2004). The idea of using a complex algorithm and a computer-aided solution to elect the representatives of Parliament opens to a number of questions. Surely such a fundamental law for democracy must be clear and transparent to all citizens and not only an optimum according to mathematicians. Moreover, the procedure must be replicable in all its steps and, above all, it must guarantee a unique solution.

Acknowledgments

My thanks to Professor Bruno Simeone and Federica Ricca for their precious suggestions and tireless interest in promoting fairer electoral systems and to Professor Michel Balinski (Ecole Polytechnique) for sharing his remarks and ideas on the Italian case.

References

Ddl Camera 2620 13 ottobre 2005 - Modifiche alle norme per l'elezione della Camera dei Deputati e del Senato della Repubblica.

L. 4 agosto 1993 n. 277 - "Nuove norme per l'elezione della Camera dei Deputati".

M. Bacharach (1970), Biproportional matrices and input/output change, Cambridge University Press, Cambridge.

M.L. Balinski, G. Demange (1989a) "Algorithms for proportional matrices in reals and integers" Mathematical Programming 45, 193-210.

M.L. Balinski and G. Demange (1989b), "An axiomatic approach to proportionality between matrices", Mathematics of Operations Research 14, 700-719.

M.L. Balinski, H.P. Young (1982), Fair Representation: Meeting the Ideal of One Man One Vote, Yale University Press, New Haven.

M.L. Balinski, V. Ramírez González (1997) "Mexican Electoral Law: 1996 version" Electoral Studies vol.16 n. 3, 329-349.

M.L. Balinski (2004) Le suffrage universel inachevé, Belin, Paris.

P. Grilli di Cortona, C. Manzi, A. Pennisi, F. Ricca B. Simeone (1999) Evaluation and Optimization of Electoral Systems, SIAM Monographs on Discrete Mathematics and Applications, Society for Industrial and Applied Mathematics (SIAM), Philadelphia.

G. Leti (1970), "La distribuzione delle tabelle della classe di Frechet", Metron, 87-119.

A. Pennisi (1999), "Disproportionality Indexes and Robust Proportional Allocation Methods", Electoral Studies, vol.17 n.1, pp.3-19.

A. Pennisi, F. Ricca, B. Simeone (2005) "Legge elettorale con paradosso" (11 Novembre 2005), "E' proprio un paradosso" (5 Dicembre 2005), www.lavoce.info.

A. Pennisi, F. Ricca, B. Simeone (2005), "Malfunzionamenti dell'allocazione biproporzionale di seggi nella riforma elettorale italiana", Dipartimento di Statistica, Probabilità e Statistiche Applicate, Technical Report n.21/2005.

F. Pukelsheim (2004), "BAZI - A Java program for proportional representation", Oberwolfach Reports 1 735-737 www.uni-augsburg.de/bazi

Current Issues of Apportionment Methods

Friedrich Pukelsheim
Institut für Mathematik, Universität Augsburg

Abstract   Three apportionment problems are addressed that are of current interest in Germany and Switzerland: the assignment of committee seats in a way that preserves the parliamentary majority-minority relation, the introduction of minimum restrictions in a two-ballot system to accomodate the direct seats won by the constituency ballots, and biproportional apportionment methods for systems with multiple districts so as to achieve proportionality between party votes as well as between district populations.

Keywords: Gentle majority clause; direct-seat restricted divisor methods; biproportional divisor methods; BAZI computer software.

1.   Introduction

Three proportional representation problems are sketched that are of practical and current interest. The first problem is to map a majority of votes into a majority of seats, encountered when the German Bundestag had to apportion sixteen committee seats. All of the methods that the Bundestag had been using so far produced a tie, assigning eight seats to the government majority and another eight to the opposition minority. A *gentle majority clause* is suggested to resolve the tie (Section 2).

The second problem concerns the election of the Bundestag deputies proper. The German Federal Electoral Law provides each voter with two ballots, a party ballot and a constituency ballot. The party ballots form the basis for a proportional apportionment of all Bundestag seats, while the constituency ballots are instrumental in identifying direct-seat winners in single-member constituencies. The Electoral Law desires to combine the two components, but actually fails to do so when setting up the operational instructions to evaluate the two ballots. Defects may evolve, the most serious – and actually fatal, in our view – defect being that more party ballots may actually cause a loss of seats. The system may thus discourage voters to cast their ballots in favor of the party of their choice! Luckily, the apportionment theory of Balinski/Young (2001) offers a remedy, by imposing minimum restrictions. *Direct-seat re-*

stricted methods evade the defects, and successfully combine the two components of the German system, of a proportional apportionment via party ballots, and of an election of persons via constituency ballots (Section 3).

The third problem considers electoral systems where the whole electoral region is subdivided into various electoral districts. We review recent work on *biproportional methods* tailored to achieve two-way proportionality, that is, proportionality among the vote counts for parties, and proportionality among the populations numbers for districts (Section 4).

2.   A Gentle Majority Clause

With the start of a legislative period, a new German Parliament [Bundestag] elects its delegates for the Bundestag-Bundesrat Conference Committee [Vermittlungsausschuss]. The Bundesrat is the assembly of the 16 states [Länder], each sending one representative into the Conference Committee. In order to be on par with the Bundesrat, the Bundestag occupies another 16 seats, apportioning them to the parliamentary factions proportional to their size. The *faction size* [Fraktiongröße] is the number of deputies belonging to the faction. In the 2002 legislative period, there were four factions, SPD, CDU/CSU, Bündnis 90/Die Grünen, and FDP, of sizes 249 : 247 : 55 : 47.

Over the years the Bundestag has familiarized itself with three apportionment methods: the divisor method with standard rounding (Webster/Sainte-Laguë/Schepers), the divisor method with rounding down (Jefferson/D'Hondt/Hagenbach-Bischoff), and the quota method with residual fit by largest remainders (Hamilton/Hare/Niemeyer). All of these methods allocate the 16 seat Bundestag delegation as $7 : 7 : 1 : 1$, entailing a tie of $8 : 8$ seats between the government majority (Social Democrats and Greens, $249 + 55 = 304$ seats), and the opposition minority (Conservatives and Liberals, $247 + 47 = 294$ seats).

To break the tie, the Bundestag majority passed a motion to proportionally apportion just 15 seats, and to directly assign the last seat to the largest faction. The resulting allocation $8 : 6 : 1 : 1$ secured a committee majority of $9 : 7$ for the government parties. Not surprisingly, the opposition minority challenged the apportionment in court. On 8 December 2004 the German Federal Constitutional Court ordered the Bundestag to reconsider the apportionment, but was otherwise vague and nebulous which constitutional principles the Bundestag was to observe when renewing its deliberations. The Court specified, though, that the procedure used ought to be "transparent, calculable, and abstract-general".

On 17 February 2005 the Bundestag Rules Committee, who was in charge of the proceedings, conducted an expert hearing. The opinion presented by us is published in Pukelsheim/Maier (2005). Our preferred option is a *gentle major-*

ity clause, consisting of two parts. First and foremost, the Bundestag attempts to select a committee size for which the divisor method with standard rounding (Webster/Sainte-Laguë/Schepers) yields an apportionment that preserves the majority-minority relation. The idea is not at all new, just codifying what already now is standard Bundestag practice. The first part, though, does not resolve the Conference Committee issue. The size of the Bundestag delegation is fixed at 16, the number of states in the federation. To evade the threatening tie, the divisor method with standard rounding (Webster/Sainte-Laguë/Schepers) needs to be amended.

The second part of the gentle majority clause comes to bear only in such cases when the first part results in a tie. Then the smallest possible majority in the committee is allocated with the government majority, thus leaving the largest possible committee minority for the opposition minority. Within each of the two groups, the seats available are apportioned using the divisor method with standard rounding (Webster/Sainte-Laguë/Schepers). For the Conference Committee, the government majority shares 9 seats in the relation 7 : 2, while the opposition minority allocates the remaining 7 seats as 6 : 1. In summary, the resulting apportionment is 7 : 6 : 2 : 1. Pukelsheim/Maier (2005) argue that the gentle majority qualifies to be transparent, calculable, and abstract-general.

3. Direct-Seat Restricted Methods

The German Federal Electoral Law provides every voter with two ballots, a *constituency ballot* [Erststimme] and a *party ballot* [Zweitstimme]. Voters mark the two ballots on a single sheet of paper where the choices for the constituency ballot are printed on the left half of the page, while the party ballot choices occupy the right half. To aid voters in distinguishing between the two halves, one is printed in blue, the other, in black.

The party ballots are the basis for the *superapportionment* [Oberzuteilung], a proportional apportionment of all 598 Bundestag seats among parties. Parties participate in the apportionment process only if they gain at least five percent of the valid party ballots. Thus the party ballots serve to run a proportional representation system with a five percent threshold, straight and simple. The system becomes more demanding when deciding who is going to fill the seats. The Law stipulates that the seats of a party are manned primarily by such candidates who, in their constituencies, won a relative majority of the constituency ballots. In other words, the objective of the constituency ballots is "to elect persons" in single-member districts. There are 299 constituencies, and hence the same number of winners of *direct seats* [Direktmandate].

The remaining 299 seats are filled with candidates from party lists. This is where the Law becomes tricky: party lists are organized by states, whence a party generally commands 16 *state lists* [Landeslisten]. Of course, the idea is that deputies have roots in the geographical region where they are elected, if not in their constituency, than at least in their state. Thus the seats that a party received in the superapportionment are proportionally broken down to its 16 state lists. Hopefully the *subapportionment* [Unterzuteilung] allocates enough seats to a party in a state, to accomodate all direct-seat winners (of that party in that state). Any additional seats are filled from the state list, usually referred to as *list seats* [Listenmandate].

There remain "exceptional" cases where a party wins more direct seats in a state than the state list receives in the subapportionment. In such cases, the direct seats stay with the party, even though the proportional allocation via super- and subapportionments does not justify that many seats. This generates additional seats, called *overhang seats* [Überhangmandate], enlarging the size of the Bundestag beyond the initial 598. While the literature sometimes speaks of "surplus seats", we stick to the experts' terminolgy deliberately coined when New Zealand adopted the German electoral system (New Zealand Electoral Commission 1986). The current Bundestag comprises 614 deputies, with 9 overhang seats for the Social Democrats and 7 for the Conservatives. Alas, the 2005 election is an "exceptional" case.

Well, since 1980 *every* Bundestag has had its overhang seats. We are using quotation marks because the "exceptional" cases occur regularly. Over the years there have been 73 overhang seats (Fehndrich 2005), of which 65 benefitted the government majority no matter whether the parties composing the majority were center, left, or right. Thus the Law grants a ninety percent chance that overhang seats boost the government majority, rather than being "misplaced" with the opposition minority.

Whoever forms the majority, it is not opportune for them to question a twist in the rules instrumental to bring them into being. The 1994 Bundestag elected Helmut Kohl Chancellor with the narrowest possible margin of one vote, his government majority providing a happy home to 12 overhang seats. Who would expect an overhang chancellor to bite the hand that voted him to power? The system defies not so much the politicians who, after all, must make the best out of a parliament as is. The challenge is up to the voters, to fight for their right to electoral equality, and to the courts, to check upon the justifiability with constitutional principles.

In essence, the malalignment of party and constituency ballots causes three defects (Pukelsheim 2000, Section II). One is overhang seats. The second is *doubly successful* votes, where the constituency ballot helps electing a deputy by circumventing her or his party's state list (because the party fails to pass the five percent hurdle, or the candidate is independent), while the party ballot still enters into the aggregation of another party list. In 2002, there were at least 270 162 voters who enjoyed the good fortune of being doubly successful (Pukelsheim 2004, page 407).

The third defect, *negative ballot weights*, is prone to prove fatal, or so we believe: more party ballots may be the cause for a party to lose a seat. The system gives rise to situations where voters are discouraged to cast their party ballots for the party of their choice!

Negative ballot weights were discussed first by Meyer (1994, page 321). The problem received some subdued press coverage, with the upshot that the electoral system entertains its oddities. Then, in the 2005 election, the defect hit all German newspapers, irritating the electorate and ridiculing the system. In the Dresden I constituency, a candidate had died shortly before the election day of September 18. This caused a shift of the election, in this constituency, to a by-election [Nachwahl] on October 2. In the main election, on September 18, the Conservatives gained four overhang seats in Sachsen state. The by-election threatened to return "too many" party votes for them, letting their proportional share grow enough to convert an overhang seat into a proportionally justified seat. The bottom line would have been the loss of one seat. The numbers speak for themselves: The Conservative voters understood, and deprived the CDU of their party ballots (Cantow/Fehndrich/Zicht 2005). The feared loss of a seat did not materialize.

Under a constitution that builds on a strict separation of powers, such as the German *Fundamental Law* [Grundgesetz], the constitutionaliy of a law is examined by the courts. The Federal Electoral Law falls under the jurisdiction of the Federal Constitutional Court. The issue of negative ballot weights was presented to the Court; surprisingly, the Court remained silent about it. With the data from the 2005 Dresden I by-election, the Court will get a chance to reconsider. The Court has otherwise upheld the Electoral Law, ruling that its commendable effort to combine the elections of persons with a proportional representation system entails the disputed defects as "necessary consequences". Here errs the Court. The defects cannot be justified as being *necessary*, in the accepted sense of the word, other than that they are consequences of the instructions in the *current* Law. There are methods evading the defects and, at the same time, coming closer to merging the two electoral principles, of electing persons and of mirroring party strenghts.

Table 1 illustrates a defect-free method, for the 2005 Bundestag election data (Schorn/Schwartzenberg 2005). The procedure is called the *direct-seat restricted divisor method with standard rounding*, and works as follows. The number d of direct seats won by a party is imposed as a minimum restriction, to make sure that enough seats are allocated to provide every constituency winner with a seat. To calculate the number of proportionally justified seats, p, the divisor method with standard rouding (Webster/Sainte-Laguë/Schepers) is used. The method divides the number of party ballots by the divisor given in Table 1, and rounds the resulting quotient in a standard fashion (down if the fractional part is below one half, and up if it is above) to obtain p. The larger of the two numbers, denoted by $d \vee p$ (read: the larger value of d or p), is the number of seats allocated. The divisors in Table 1 are such that the seats apportioned exhaust the seats available.

For example, in the superapportionment the 16 194 665 party ballots of the SPD are divided by 76 000. The resulting quotient 213.1 is rounded to $p =$ 213. Since this exceeds the number of direct seats, $d = 145$, the SPD is eligible to 213 seats, on the federal level. In the subapportionment, the 213 seats are broken down to the 16 SPD state lists. The divisor used is 80 000, shown at the bottom of the column. The SPD in Sachsen-Anhalt (ST) won $d = 10$ direct seats, but received just $p = 6$ proportionally justified seats. Formerly, the difference would have generated four overhang seats. With the direct-seat restricted method, the larger of the two numbers applies, 10. For the SPD, the direct seat component dominates in five states (HH, BB, ST, TH, SL), in two states the tally is balanced (MV, HB), and in the other nine the proportionally justified seats are effective.

4.   Biproportional Methods

The subdivision of a single large electoral region into various smaller *electoral districts* is an ubiquitous topic. The German Electoral Law, dealing with sixteen states, provides just one way of handling the issue. Another well-established approach allocates the total number of seats to the electoral districts proportionally to population counts, some time during the legislative period. With the seat numbers for each district prespecified, the votes are then evaluated separately in each district. This is the system that was in use in the Canton of Zurich, Switzerland. Due to population mobility, however, some districts shrunk to as few seats as two, in the presence of some seven and more parties competing. Naturally, the idea of proportionality must fail when apportioning just two seats among many competitors. This provided the motivation to switch to a *biproportional method*.

Biproportional apportionment methods were introduced into the literature by Balinski/Demange (1989a,b). Balinski (2002) applied the method to Mexico, in a popular science article that I translated into German. Shortly afterwards Christian Schuhmacher, from the Zurich Justice and Interior Department, hit upon the Augsburg Bazi group in the Internet. Together, we adopted Balinski's idea to the Zurich situation (Pukelsheim/Schuhmacher 2004). The *new Zurich apportionment procedure* [Neues Zürcher Zuteilungsverfahren, NZZ] had its world debut with the Zurich City Parliament election on 12 February 2006 (Balinski/Pukelsheim 2006a,b).

Table 1. Election of the sixteenth German Bundestag on 18 September 2005, direct-seat restricted divisor method with standard rounding.

	SPD	CDU	FDP	Die Linke	Grüne	CSU
	Superapportionment of the 598 Bundestag seats to parties (Divisor = 76 000)					
	16 194 665	13 136 740	4 648 144	4 118 194	3 838 326	3 494 309
	145 ∨ 213 = 213	106 ∨ 173 = 173	0 ∨ 61 = 61	3 ∨ 54 = 54	1 ∨ 51 = 51	44 ∨ 46 = 46
	Subapportionment of overall party seats to state lists (na = no list submitted)					
SH	655 361	624 510	173 320	78 755	144 712	na
	5 ∨ 8 = 8	6 ∨ 8 = 8	0 ∨ 2 = 2	0 ∨ 1 = 1	0 ∨ 2 = 2	
MV	314 830	293 316	62 049	234 702	39 379	na
	4 ∨ 4 = 4	3 ∨ 4 = 4	0 ∨ 1 = 1	0 ∨ 3 = 3	0 ∨ 1 = 1	
HH	365 546	272 418	84 593	59 463	140 751	na
	6 ∨ 5 = 6	0 ∨ 3 = 3	0 ∨ 1 = 1	0 ∨ 1 = 1	0 ∨ 2 = 2	
NI	2 058 174	1 599 947	426 341	205 200	354 853	na
	25 ∨ 26 = 16	4 ∨ 20 = 20	0 ∨ 6 = 6	0 ∨ 3 = 3	0 ∨ 5 = 5	
HB	155 366	82 389	29 329	30 570	51 600	na
	2 ∨ 2 = 2	0 ∨ 1 = 1	0 ∨ 0 = 0	0 ∨ 0 = 0	0 ∨ 1 = 1	
BB	561 689	322 400	107 736	416 359	80 253	na
	10 ∨ 7 = 10	0 ∨ 4 = 4	0 ∨ 1 = 1	0 ∨ 5 = 5	0 ∨ 1 = 1	
ST	474 909	357 663	117 155	385 422	59 146	na
	10 ∨ 10 = 10	0 ∨ 5 = 5	0 ∨ 2 = 2	0 ∨ 5 = 5	0 ∨ 1 = 1	
BE	637 674	408 715	152 157	303 630	254 546	na
	7 ∨ 8 = 8	1 ∨ 5 = 5	0 ∨ 2 = 2	3 ∨ 4 = 4	1 ∨ 3 = 3	
NW	4 096 112	3 524 351	1 024 924 0 13	529 967	782 551	na
	40 ∨ 51 = 51	24 ∨ 44 = 44	24 ∨ 44 = 44	0 ∨ 7 = 7	0 ∨ 10 = 10	
SN	649 807	795 316	269 623	603 824	126 850	na
	3 ∨ 8 = 8	14 ∨ 10 = 14	0 ∨ 4 = 4	0 ∨ 8 = 8	0 ∨ 2 = 2	
HE	1 197 762	1 131 496	392 123	178 913	340 288	na
	13 ∨ 15 = 15	8 ∨ 14 = 14	0 ∨ 5 = 5	0 ∨ 2 = 2	0 ∨ 5 = 5	
TH	432 778	372 435	115 009	378 340	69 976	na
	6 ∨ 5 = 6	3 ∨ 5 = 5	0 ∨ 1 = 1	0 ∨ 5 = 5	0 ∨ 1 = 1	
RP	822 074	877 632	278 945	132 154	172 900	na
	5 ∨ 10 = 10	10 ∨ 11 = 11	0 ∨ 4 = 4	0 ∨ 2 = 2	0 ∨ 2 = 2	
BY	1 806 548	na	673 817	244 701	559 941	3 494 309
	1 ∨ 23 = 23		0 ∨ 9 = 9	0 ∨ 3 = 3	0 ∨ 7 = 7	44 ∨ 46 = 46
BW	1 754 834	2 283 085	693 835	219 105	623 091	na
	4 ∨ 22 = 22	33 ∨ 29 = 33	0 ∨ 9 = 9	0 ∨ 3 = 3	0 ∨ 8 = 8	
SL	211 201	191 067	47 188	117 089	37 489	na
	4 ∨ 3 = 4	0 ∨ 2 = 2	0 ∨ 1 = 1	0 ∨ 2 = 2	0 ∨ 0 = 0	
Divisor	*80 000*	*79 300*	*77 000*	*77 000*	*75 000*	*76 000*

SH Schleswig-Holstein	HB Bremen	NW Nordrhein-Westfalen	RP Rheinland-Pfalz
MV Mecklenburg-Vorpommern	BB Brandenburg	SN Sachsen	BY Bayern
HH Hamburg	ST Sachsen-Anhalt	HE Hessen	BW Baden-Württemberg
NI Niedersachsen	BE Berlin	TH Thüringen	SL Saarland

The seats apportioned are written as $d \vee p$, that is, the larger value of d or p, where d is the count of direct seats won and p is number of proportionally justified seats. In the superapportionment, the SPD entry 145 ∨ 213 = 213 means that the party won 145 direct seats, while its party ballots justify 213 seats; the larger number prevails, 213.

Table 2. Biproportional divisor method with standard rounding, retrospectively applied to the 2002 Zurich City Parliament election.

		SP	SVP	FDP	Grüne	CVP	SenL	AL	*City divisor*
Electorate support		33287	17753	15307	8299	6072	3475	3223	*710*
				Biproportional apportionment of overall party and district lists					
	125	*47*	*25*	*22*	*12*	*9*	*5*	*5*	*District divisor*
"1+2"	*12*	42192-4	20508-2	28956-3	12960-2	7668-1	2964-0	2208-0	*9600*
"3"	*16*	68219-6	28897-3	16992-2	13752-2	8619-1	5428-1	8040-1	*10400*
"4+5"	*13*	40339-6	9854-1	7358-1	11271-2	6071-1	1781-0	12220-2	*7000*
"6"	*10*	36257-4	13491-2	14874-2	9556-1	4708-1	3592-0	2797-0	*8000*
"7+8"	*17*	84456-5	41191-2	74018-5	32963-2	16456-1	8245-1	6987-1	*16480*
"9"	*16*	58119-6	43585-5	20258-2	11681-1	15130-1	7717-1	3684-0	*9500*
"10"	*12*	49241-5	25620-2	24797-3	10621-1	7762-1	5351-0	4355-0	*10706*
"11"	*19*	77998-7	63333-5	30541-3	14643-1	18027-1	12088-1	4685-1	*11515.5*
"12"	*10*	19700-4	15159-3	4861-1	2105-0	4462-1	3438-1	650-0	*5000*
Party divisor		*1.022*	*1*	*0.9*	*0.87*	*1.08*	*1*	*0.81366*	

The table entries p-s list party votes p and seat numbers s. To obtain s, party votes p are divided by the associated district and party divisors, and then rounded. In District "1+2", party SP wins $p = 42192$ votes and gets $s = 4$ seats, since $p/(9600 \times 1.022) = 4.3 \searrow 4$. The divisors (right and bottom, in italics) are such that the prespecified district seats and the overall party seats (left and top, in italics) are met exactly. The overall party seats result from the superapportionment, on the basis of electorate supports.

Table 2 shows the method at work in a hypothetical, restrospective evaluation of the past 2002 election data. In order to participate in the apportionment process, the five percent threshold must be passed in at least one district. In 2002, this would have left seven parties. The first step then is the *superapportionment*, the apportionment of all 125 parliament seats among parties, proportionally to their electorate support. This step responds to the constitutional demand that all voters contribute to the electoral outcome equally. Other than with the former system of separate district evaluations, it no longer matters whether voters cast their ballots in districts that are large or small. The second step is the *subapportionment*: The overall party seats are handed down to the districts, while verifying the prespecified district totals. Mathematics guarantees that, when a biproportional method is used, the resulting apportionment is unique (up to ties).

A complication arises since a Zurich voter is provided with as many ballots as the district has seats to fill. Thus voters in District "1+2" command 12 ballots, in District "3" they have 16, etc. The ballots may be split among parties, and cumulated. The resulting counts are called *party ballots* [Parteistimmen]; these are the raw data returned from the polling stations. The districtwise party ballots need to be aggregated across the whole electoral region. To this end, party ballots are divided by the district magnitude and rounded, yielding the *district support* [Wahlkreis-Wählerzahl] of a party. The sum of the district supports is called the *electorate support* [(Kanton-)Wählerzahl] of a party, indicating how many voters back the party across the whole electoral region. Since the conversion to support quantities adjusts for the distinct number of ballots handed out in a district, every voter contributes to the superapportionment in an equal manner.

In Table 2, the SP enjoys in District "1+2" a district support of $42192/12 = 3516$, while in District "3" the support is $68219/16 = 4263.7 \nearrow 4264$. The seven parties participating in the apportionment process turn out to win electorate supports of $33287 : 17753 : 15307 : 8299 : 6072 : 3475 : 3223$. Using the divisor method with standard rounding (Webster/Sainte-Laguë/Schepers), the superapportionment allocates the 125 seats according to $47 : 25 : 22 : 12 : 9 : 5 : 5$ (city divisor 710).

The subapportionment employs the *biproportional method with standard rounding*. It achieves a two-way proportionality, while verifying the prespecified district magnitudes as well as exhausting the overall party seats just calculated. The restrictions form the left and top borders of Table 2, typeset in italics. The method aims at proportionality among the party votes that form the table body. Two sets of divisors come into play, *district divisors* and *party divisors*, bordering Table 2 on the right and at the bottom (in italics).

The method divides the party votes by the associated district and party divisors, and rounds the resulting quotient in a standard fashion to obtain the seat number. For instance, the SP in District "1+2" receives $42192/(9600 \times 1.022) = 4.3 \searrow 3$ seats. The same district divisor is used for the vote counts of all parties, in any given district, thus treating parties districtwise equally. Similarly, the same party divisor is applied to the vote counts in all districts, for any given party, again honoring the proportionality principle. The biproportional apportionment is *coherent*, in that it fairly approximates the ideal shares of seats a party may claim when contesting individual seats (Balinski/Pukelsheim 2006b).

References

M. Balinski (2002). Une "dose" de proportionnelle – Le systéme électoral mexicain. *Pour la science*, April 2002, 58–59. [German translation: Verhältniswahlrecht häppchenweise – Wahlen in Mexiko. *Spektrum der Wissenschaft*, October 2002, 72–74.]

M. Balinski / G. Demange (1989a). An axiomatic approach to proportionality between matrices. *Mathematics of Operations Research*, 14, 700–719.

M. Balinski / G. Demange (1989b). Algorithms for proportional matrices in reals and integers. *Mathematical Programming*, 45, 193–210.

M. Balinski / F. Pukelsheim (2006a). Matrices and politics. In: *Matrices and Statistics. Festschrift for Tarmo Pukkila on his 60th Birthday.* Editors E.P. Liski / J. Isotalo / J. Niemelä / S. Puntanen / G.P.H. Styan, Department of Mathematics, Statistics, and Philosophy, University of Tampere, 233–242.

M. Balinski / F. Pukelsheim (2006b). A double dose of proportionality: Zürich's new electoral law. Typescript.

M. Balinski / H.P. Young (2001). *Fair Representation – Meeting the Ideal of One Man, One Vote. Second Edition.* Washington, DC.

M. Cantow / M. Fehndrich / W. Zicht (2005).
www.wahlrecht.de/bundestag/2005/nachwahl-dresden-spezial.html

M. Fehndrich (2005). Historisches zu Überhangmandaten.
www.wahlrecht.de/ueberhang/ueberhist.html

H. Meyer (1994). Der Überhang und anderes Unterhaltsames aus Anlaß der Bundestagswahl 1994. *Kritische Vierteljahresschrift für Gesetzgebung und Rechtswissenschaft*, 77, 312–362.

New Zealand Electoral Commission (1986). *Towards a Better Democracy – Report of the Royal Commission on the Electoral System.* Editor Electoral Commission. Wellington, 1986. [Reprint: Wellington, 1997.]

F. Pukelsheim (2000). Mandatszuteilungen bei Verhältniswahlen: Vertretungsgewichte der Mandate. *Kritische Vierteljahresschrift für Gesetzgebung und Rechtswissenschaft*, 83, 76–103.

F. Pukelsheim (2004). Erfolgswertgleichheit der Wählerstimmen zwischen Anspruch und Wirklichkeit. *Die Öffentliche Verwaltung*, 57, 405–413.

F. Pukelsheim / S. Maier (2005). Eine schonende Mehrheitsklausel für die Zuteilung von Ausschusssitzen. *Zeitschrift für Parlamentsfragen*, 36, 763–772. [English translation: A gentle majority clause for the apportionment of committee seats. This volume, 177–188.]

F. Pukelsheim / C. Schuhmacher (2004). Das neue Zürcher Zuteilungsverfahren für Parlamentswahlen. *Aktuelle Juristische Praxis – Pratique Juridique Actuelle*, 5, 505–522.

K. Schorn / M. von Schwartzenberg (2005). Endgültiges Ergebnis der Wahl zum 16. Deutschen Bundestag am 18. September 2005. *Statistisches Bundesamt, Wirtschaft und Statistik*, 11/2005, 1153–1167.

A Gentle Majority Clause for the Apportionment of Committee Seats

Friedrich Pukelsheim, Sebastian Maier

Institut für Mathematik, Universität Augsburg

Abstract   The divisor method with standard rounding (Sainte-Laguë/Schepers) is amended by a gentle majority clause, in order to map the government majority in parliament into a seat majority in committees.

Keywords: Divisor method with standard rounding (Sainte-Laguë/Schepers); D'Hondt; Hill; Hare/Niemeyer; Adams; Condorcet.

1.   Itio in Partes

We address the problem of how to constitute legislative committees while attempting to reconcile two objectives that sometimes conflict in closely divided legistlatures: representing parties proportional to their seats in the legislature, and maintaining control of the committee by the party or coalition that enjoys a majority in the legislature. The problem arises in many legislatures at the national, state, and municipal levels. A notable recent instance occurred for the German Bundestag, and led to the December 2004 decision of the German Federal Constitutional Court, concerning the composition of the 16 seat Bundestag delegation in the Bundestag-Bundesrat Conference Committee.[1] On 17 February 2005, the Rules Committee of the German Bundestag conducted an expert hearing to elucidate the Court's decision. The present paper is the solution that the authors recommended to the Bundestag, and closely follows their testimony.[2]

---

[1]Decision of 8 December 2004 (Az. 2 BvE 3/02), here quoted using the marginal running numbers (Rn.) of the Internet publication www.bverfg.de/entscheidungen/es20041208_2bve000302.html. — As far as the German Bundestag-Bundesrat Conference Committee is concerned, a corrective action violating proportionality in order to preserve the government majority is considered inadmissible by J. Masing, who finds the contrary conclusion in the decision of the German Federal Constitutional Court inconsistent and nebulous, see Section C.I.3 of his commentary on Art. 77 GG in Mangoldt/Klein/Starck (2005). — See also Kämmerer (2003), Lovens (2003), Stein (2003), Lang (2005).

[2]Pukelsheim/Maier (2005). See also Meyer (2005).

Our proposal of a "gentle majority clause" builds on historic precedence. Inspired by the Pax Augustana of 1455, proclaimed in Augsburg some 450 years ago, the peace of Westphalia of 1648 codified constitutional clauses backing a peaceful coexistence of the two dominating Christian confessions. This included the procedural parity of an *itio in partes*.[3] The *splitting into parts* guaranteed an equal treatment of two unequal groups when the preservation of the mutual identities was considered essential for the whole body. In the then confessional age, the parts were the *Corpus Catholicorum* and the *Corpus Evangelicorum*. In today's democracies, the two groups are majority and minority.

For the expert hearing, the Bundestag Rules Committee compiled a catalogue of five questions. Question 1 concerns the constitutionality of obtaining a mirror image, or of preserving a parliamentary majority. Questions 3–5 aim at procedural and other legislative consequences. Mathematics cannot contribute to these questions. Question 2, adressed in the sequel, asks which operational options are available under the premise that a preservation of the parliamentary majority does conform with the Constitution:

2. *If it is constitutionally legitimate to preserve the majority,*

      a) which measures (for example seat numbers of the factions; relation between majority and opposition),

      b) which procedural possibilities (for example combination of one of the usual apportionment procedures with a correction factor; choice of a hitherto not practiced, but majority preserving apportionment procedure, other alternatives) and

      c) which changes to the rules and standing orders of the Bundestag would be called for in order to achieve a "gentle balance"?

The notion of a "gentle balance" [schonender Ausgleich] is taken from the Court decision.[4] However, we find the wording "balance" somewhat besides the point, and instead speak of *majority clauses*.

2.   A Gentle Majority Clause

On 30 October 2002, right at the beginning of the legislative period, the Bundestag passed a motion on how to apportion committee seats.[5] The motion comprised two parts, of which Part 1) poses no particular problems:

1) The number of committee seats apportioned to a faction and the sequence of the allocation of chairpersons, of the Steering Committee and

---

[3]Heckel (1978), Burkhardt (1998). — The Court decision (Rn. 76) refers to the *itio in partes* in US Senate-House conference committees, see Riescher/Ruß/Haas (2000, page 39), or in the Internet www.house.gov/rules/98-382.pdf.
[4]BVerfGE 2 BvE 3/02, Rn. 64, 77, 84, 86. Dissenting: Rn. 112.
[5]BVerfGE 2 BvE 3/02, Rn. 8–10.

of the other committees of the German Bundestag, are determined by means of the procedure of mathematical proportions (Sainte-Laguë/Schepers), unless the Bundestag decides otherwise.

The same procedure is used for the apportionment of seats to other parliamentary bodys, unless a different procedure is stipulated by law.

Rather than using the term "procedure of mathematical proportions (Sainte-Laguë/Schepers)", we speak of the *divisor method with standard rounding (Sainte-Laguë/Schepers)*, thus providing some guidance about how the seat apportionments are calculated.[6] For example, for a delegation of size 16 the current faction sizes 249 : 247 : 55 : 47 result in an apportionment of 7 : 7 : 1 : 1 seats (divisor 37). Hence the government majority and the opposition minority are tied, with 8 seats each. Part 2) of the Bundestag motion serves as a tie breaking rule, to be called the *prevailing majority clause*:

2) If the parliamentary majority is not preserved, the method of D'Hondt is used. If this method also fails to preserve the parliamentary majority, the method of Sainte-Laguë/Schepers is used with the amendment that the number of seats to be apportioned is reduced by one and that the remaining seat is given to the largest faction.

For a delegation of size 16, the second sentence of Part 2) applies. Thus 15 seats are apportioned using the divisor method with standard rounding (Sainte-Laguë/Schepers), giving in an intermediate allocation of 7 : 6 : 1 : 1 seats (divisor 38.2). The sixteenth seat is given to the largest faction, resulting in a final apportionment of 8 : 6 : 1 : 1 seats.

The Court decision seems to indicate, or so we believe, that the prevailing majority clause secures a somewhat questionable advantage for the largest faction.[7] From the viewpoint of mathematics, the prevailing majority clause simply lacks general applicability.[8] The following proposal, to be called the *gentle majority clause*, applies quite generally:

---

[6]In the Data Handbook of the German Bundestag, the method is called the "Proportional procedure (of Sainte-Laguë/Schepers)", see Schindler (1999, Volume II, page 2085). The method is attributed to *Daniel Webster* (1782–1813), see Balinski/Young (2001). — *André Sainte-Laguë* [sɛ̃t laˈgy] (1882–1950) was professor of *Mathématiques générales en vue des applications* with the *Conservatoire national des arts et métiers* in Paris. *Hans Schepers* (*1928) was Head of the Data Processing Group of the scientific staff of the German Bundestag (Pukelsheim 2002). Sainte-Laguë was not a saint, whence it is inappropriate to shorten his name to "St. Laguë" or "Ste. Laguë" — Sample calculation: The quotient $249/37 = 6.7$ is rounded in standard fashion to 7, as is $247/37 = 6.7 \nearrow 7$, and $55/37 = 1.49 \searrow 1$, as well as $47/37 = 1.3 \searrow 1$. The divisor 37 is indicative of 37 deputies being represented by one (up to rounding) delegate. — The government majority was composed by SPD (249 deputies) and Greens (55), the opposition minority by CDU/CSU (247) and Liberals (47).

[7]BVerfGE 2 BvE 3/02, Rn. 83, 85.

[8]For example, a transfer of ten opposition seats from FDP to CDU/CSU turns the faction sizes into 249 : 257 : 55 : 37, making CDU/CSU the largest faction to be awarded the bonus seat. Hence the intermediate allocation 6 : 7 : 1 : 1 (divisor 39) leads to the final apportionment of 6 : 8 : 1 : 1 seats. Although government majority and opposition minority stay put at 304 : 294 deputies, the prevailing majority clause produces a *majority reversal* of 7 : 9 seats in the committee.

2) If the government majority is to be preserved, then first and foremost it is attempted to achieve the goal by selecting an appropriate committee size. Otherwise, the smallest possible committee majority is apportioned among the factions composing the government majority, while the remaining committee seats are apportioned among the remaining factions; both apportionments are calculated by means of the divisor method with standard rounding (Sainte-Laguë/Schepers).

Applying the gentle majority clause to a delegation of size 16, the government majority gets allocated 9 seats and the opposition minority 7 seats. The two factions forming the government majority have 249 : 55 deputies whence they allocate their 9 seats into 7 : 2 (divisor 35). The opposition minority, with 247 : 47 parliamentary seats, share their 7 seats as 6 : 1 (divisor 38.2). In summary, the 16 seats are apportioned into 7 : 6 : 2 : 1.

In the first sentence, the gentle majority clause honors standard practice of the Bundestag. If feasible, the best way-out is to select a committee size evading a tie. The second sentence comes into play only when this road is blocked. In these exceptional cases, the committee is split into a majority part and a minority part, applying the divisor method with standard rounding (Sainte-Laguë/Schepers) to the two groups separately.[9]

3.   Transparency, Calculability, and Abstract Generality

The Federal Constitutional Court demands of the Bundestag to formulate deviations from the majority principle in a transparent, calculable, and abstract-general manner.[10] As far as deviations from the majority principle are concerned, a transition from the prevailing majority clause to the gentle majority clause would not introduce any changes. The reservation at the end of the first paragraph in Part 1) of the Bundestag motion allows to enact other procedures for particular cases (Children's Commission, Conference Committee etc.), if so desired.

However, we find it appropriate to emphasize that a transition to the gentle majority clause generates deviations from the mirror image principle that conform with the Court's standards. Indeed, Part 2) of the gentle majority clause is transparent and explicit. It maintains a global mirror image as long as possible. The whole committee splits into a majority group and a minority group only when necessary. But even then the mirror image principle is followed as much as possible, by properly apportioning seats separately within each of the two groups. Moreover, the gentle majority clause is calculable and

---

[9]We emphasize that the same method is applied with and without a split, and, if split, within either group. A paradoxical seat transfer triggered by a change of apportionment methods is reported by M. Fehndrich, on the Internet site (www.wahlrecht.de/systemfehler/zweiverfahren.html).
[10]BVerfGE 2 BvE 3/02, Rn. 86.

abstract-general. Table 1 illustrates the usage of the gentle majority clause, with committee sizes from 1 up to 45. Every committee size preserves the government majority, requiring a split into majority and minority parts in the fifteen rows marked with a star $*$.[11]

The consistency of Table 1 is remarkable: There are no backward jumps![12] The seat apportionments for the majority group stay the same or increase, but never decrease; the same applies to the seat apportionments of the minority. Since divisor methods are coherent, a merger of the two within-group apportionments yields the same global apportionment that is obtained from a one-step calculation (without a split into majority and minority groups) whenever the latter is such that the majority is preserved.[13]

4.   Success-Value Equality of the Deputies' Votes

Electoral systems should be judged not so much on the basis of such executive attributes as transparency, calculability, and abstract generality. Instead the judgment should focus on the question of whether the system satisfies the principle of electoral equality. The decision of the German Federal Constitutional Court touches this issue only in passing.[14]

The apportionment of committee seats involves three groups of actors that each can put forward a constitutional claim to equality: The deputies, the factions, and the committee members. From a mathematical viewpoint there is a structural similarity for the transitions, from Bundestag deputies to committee members via the apportionment method laid down in the Bundestag rules, and from voters to Bundestag deputies via the electoral system set forth in the Federal Electoral Law. For the Electoral Law, the Federal Constitutional Court interprets the abstract principle of electoral equality as "success-values equality" [Erfolgswertgleichheit] of the voters' ballots.

In the same vein, the problem of apportioning committee seats calls for an equal success-value of the deputies who are being represented in the commit-

---

[11] In a newly convening Bundestag it would then suffice to work with this one table of seat apportionments, only, rather than with the three tables used up to now: a first table with the Sainte-Laguë/Schepers apportionments, a second table with D'Hondt apportionments, and a third table with Hare/Niemeyer apportionments.
[12] Called "illogical jumps" in the Handbook of the German Bundestag, see Schindler (1999, Volume II, page 2084).
[13] Balinski (2004a, page 196; 2004b). Balinski/Young (2001, page 141) speak of *uniformity* in place of *coherence.* – Let $M = 1, 2, 3 \ldots , 45$ denote the committee size. The smallest possible majority then comprises $(M + 1)/2$ seats when M is odd, and $(M + 2)/2$ seats when M is even. Hence the minority is assigned $(M - 1)/2$ or $(M - 2)/2$ seats according as M is odd or even:

Committee size:	1	2	3	4	5	6	7	8	9	10	11	12	13	14	15	...	45	M
Majority:	1	2	2	3	3	4	4	5	5	6	6	7	7	8	8	...	23	$\lceil (M + 1)/2 \rceil$
Minority:	0	0	1	1	2	2	3	3	4	4	5	5	6	6	7	...	22	$\lfloor (M - 1)/2 \rfloor$

Thus the "next" seat alternates between majority and minority, in the range of seats considered.
[14] BVerfGE 2 BvE 3/02, Rn. 82. Dissenting: Rn. 107–129.

Table 1:　Apportionment of committee seats
using the gentle majority clause[a]

Seats	SPD	CDU/CSU	B90/Die Grünen	FDP	Divisor(s)
1	1	0	0	0	496
*2	2	0	0	0	165; 496
3	2	1	0	0	165
*4	2	1	1	0	100; 165
5	2	2	1	0	100
6	3	2	1	0	99
7	3	3	1	0	96
*8	4	3	1	0	71; 96
9	4	3	1	1	71
*10	5	3	1	1	55; 71
11	5	4	1	1	55
*12	6	4	1	1	45; 55
13	6	5	1	1	45
*14	7	5	1	1	38.2; 45
15	7	6	1	1	38.2
*16	7	6	2	1	35; 38.2
17	7	7	2	1	35
18	8	7	2	1	33
19	8	8	2	1	32
*20	9	8	2	1	29.2; 32
21	9	8	2	2	29.2
*22	10	8	2	2	26.1; 29.2
23	10	9	2	2	26.1
*24	11	9	2	2	23.6; 26.1
25	11	10	2	2	23.6
*26	11	10	3	2	21.8; 23.6
27	11	11	3	2	21.8
28	12	11	3	2	21.6
29	12	12	3	2	20
30	13	12	3	2	19.8
31	13	13	3	2	19
*32	14	13	3	2	18.4; 19
33	14	13	3	3	18.4
*34	15	13	3	3	17.1; 18.4
35	15	14	3	3	17.1
*36	16	14	3	3	16; 17.1
37	16	15	3	3	16
*38	16	15	4	3	15.4; 16
39	16	16	4	3	15.4
40	17	16	4	3	15
41	17	17	4	3	14.6
42	18	17	4	3	14.2
43	18	18	4	3	14
44	19	18	4	3	13.44
45	19	18	4	4	13.4

[a]on the basis of faction sizes on 1 February 2005: SPD 249, CDU/CSU 247, Bündnis 90/Die Grünen 55, FDP 47.

All apportionments are calculated using the divisor method with standard rounding (Sainte-Laguë/Schepers). In lines marked * two separate calculations are carried out, one for the majority group and one for the minority group.

Sample calculation for committee size *16: The majority divisor 35 yields $249/35 = 7.1 \searrow 7$ and $55/35 = 1.6 \nearrow 2$. The minority divisor 38.2 leads to $247/38.2 = 6.47 \searrow 6$ and $47/38.2 = 1.2 \searrow 1$.

Sample calculation for committee size 18: The divisor 33 gives $249/33 = 7.55 \nearrow 8$ and $247/33 = 7.48 \searrow 7$ and $55/33 = 1.67 \nearrow 2$ and $47/33 = 1.4 \searrow 1$.

tee. Our preferred proposal of a gentle majority clause builds on the divisor method with standard rounding (Sainte-Laguë/Schepers). The reason is that this method produces seat apportionments that are in an exceptional harmony with the principle of success-value equality whence, in this very specific sense, the method is superior to other competing apportionment methods.[15]

The gentle majority clause is our preferred proposal because it does away with an ecclectic multitude of apportionment methods, and builds solely on the success-value oriented divisor method with standard rounding (Sainte-Laguë/ Schepers).

5. Preservation of the Majority by Means of D'Hondt

In the remaining sections we discuss other possibilities to respond to the Rules Committee's Question 2.[16]

For an appraisal of the following alternatives we recall that the prevailing majority clause, in its Part 2), resorts to the divisor method with rounding down (D'Hondt) for the reason that this method is known to be biased, in favor of larger participants and at the expense of smaller participants. These seat biases do not materialize every time the method is applied, but become clearly visible in repeated applications. As it happens, for the problem under discussion, with faction sizes $249 : 247 : 55 : 47$ and committee size 16, the D'Hondt method results in the already familiar tie $7 : 7 : 1 : 1$ (divisor 33).[17]

---

[15] See Pukelsheim (2000a, b, c).

[16] It would also be conceivable to apply the German Federal Electoral Law which (as of this writing) employs the quota method with residual fit by largest remainders (Hare/Niemeyer). — *Thomas Hare* (1806–1891) was a barrister and *Inspector of Charities* in London. *Horst F. Niemeyer* (*1931) is Professor emeritus for Mathematics with the Rheinisch-Westfälische Technische Hochschule Aachen. — The Federal Electoral Law (BWahlG) contains in its §6(3) a majority clause. Its constitutionality has been confirmed by NdsStGHE 1 (1978, pages 335–372). To apply this clause to a committee of size 16, the calculations are as follows. The faction sizes $249 : 247 : 55 : 47$ are divided by the quota $598/16$ and result in the ideal shares $6.66 : 6.61 : 1.47 : 1.26$. This gives rise to the main apportionment $6 : 6 : 1 : 1$, leaving two residual seats. According to §6(3) BWahlG, the majority is preserved by appropriately assigning the residual seats, leading to the final apportionment $7 : 6 : 2 : 1$. — Alternatively, one could carry out the calculations in two steps, with a split into two parts. Considering the majority and minority groups, of $304 : 294$ deputies, their ideal shares are $8.13 : 7.87$ and lead to the main apportionment $8 : 7$. According to §6(3) BWahlG, the remaining residual seat is allocated with the majority group, whence the two groups end up with $9 : 7$ seats. The sub-apportionments of the 9 seats within the majority, and of the 7 seats within the minority yield the same final apportionment $7 : 6 : 2 : 1$ as before. — For committees of size 8 and 12 either way leads to the apportionments $4 : 3 : 1 : 0$ and $6 : 4 : 1 : 1$, which coincide with those given in Table 1.

[17] In the Data Handbook of the Bundestag the method is called "Höchstzahlverfahren (nach D'Hondt)", see Schindler (1999, Volume II, page 2083). — *Victor D'Hondt* (1841–1901) was Professor for Civil Law and Financial Law with the University of Gent. He himself and his contemporaries spelled his name with a capital initial "D", librarians file the name under the letter "H". In Switzerland, the method is named after *Eduard Hagenbach-Bischoff* (1833-1910), Professor of Physics with the University of Basel. — Sample calculation: After subdivision by the divisor, all resulting quotients are rounded down: $249/33 = 7.5 \searrow 7$, and $247/33 = 7.5 \searrow 7$, and $55/33 = 1.7 \searrow 1$, and $47/33 = 1.4 \searrow 1$. — For the succession of apportionment methods the Bundestag has used so far, from D'Hondt via Hare/Niemeyer (from 1970 on) to Sainte-Laguë/Schepers (from 1980 on), see Fromme (1970), and Schindler (1999, Volume II, page 2081–

Of the fifteen tied rows in Table 1, ten persist under the divisor method with rounding down, while five ties are resolved. For instance, in a committee of size 32, the divisor method with standard rounding (Sainte-Laguë/Schepers) leads to the tie 13 : 13 : 3 : 3 (divisor 18.7). In contrast, the divisor method with rounding down (D'Hondt) transfers a seat from the smallest to the largest participant and yields 14 : 13 : 3 : 2 (divisor 17.8), which is the same apportionment resulting from the gentle majority clause in Table 1.

We may summarize the effects of the divisor method with rounding down (D'Hondt) as follows. At best it produces the same result as the gentle majority clause. Otherwise, it may preserve the majority without, however, securing within the majority and minority groups success-values as balanced as those coming with the gentle majority clause. And there is the third possibility that the method re-produces the tie it was suppose to resolve.

6.   A Brutal Majority Clause

Technically, a split into majority and minority groups can also be implemented with the divisor method with rounding down (D'Hondt). The government majority, commanding 249 : 55 deputies, would share their 9 committee seats in the proportion 8 : 1 (divisor 30). The opposition minority, with 247 : 47 Bundestag seats, would be allocated 6 : 1 committee seats (divisor 40). The resulting apportionment is 8 : 6 : 1 : 1, which is the seat allocation contested in Court. From our point of view as mathematicians, this majority clause is brutal and hard to defend. The split into majority and minority groups is aggravated by the seat biases inherent in the divisor method with rounding down (D'Hondt). The brutal majority clause comes with a greater deviation from proportionality than is needed for a gentle, minimal intervention.[18]

---

4). — When the D'Hondt method is applied to four participants, the largest participant can expect an advantage of +0.5 seat fractions, the second largest +0.1 fractions. To even out these advantages, the third participant misses its ideal share on the average by −0.2 fractions of a seat, the smallest participant by −0.4. See Schuster/Pukelsheim/Drton/Draper (2003, page 663).

[18]The German Federal Constitutional Court might well (presumably, at present) judge the brutal majority clause to be constitutional. In fact, the Court puts the divisor method with rounding down (D'Hondt) on a par with the divisor method with standard rounding (Sainte-Laguë/Schepers), even though the D'Hondt method exhibits noticable seat biases, while the Sainte-Laguë method is exceptionally concordant with the Court's imperative of success-value equality. Other German courts circumnavigate the shallowness in the decisions of the Federal Constitutional Court, by stating that the D'Hondt method is generally admissible, but then overruling its specific apportionment results as unlawful: due to multiple applications in separate electoral districts (BayVerfGHE 45, pages 12–23, 54–67, 85–89), due to a misuse of list combinations (BVerwG Az. 8 C 18.03 of 10 December 2003), due to a deviation from the ideal shares (BayVerwGH Az. 4 BV 03.117 and Az. 4 BV 03.1159 of 17 March 2004). We take this casuistry as a first evidence that the legal viewpoint is changing, as is implied by the State Court for the Land Baden-Württemberg (decision of 24 March 2003, Az. GR 3/01, Section B.III.2.b). A second evidence is the fact that appelants who lost their court case did not appeal to the top Federal courts although the contested facts were *not* unconstitutional (explicit: page 192 in BayVerfGH 47, 1994, 84-194; implicit: page 283 in BVerfGE 96, 1998, 264-288). A revision to the top Federal courts may induce these courts to turn to the Federal Constitutional Court for

7.   Preservation of the Majority by Means of Hill et al.

If the divisor method with rounding down (D'Hondt) induces a tie-break, it does so for the reason that a seat of a smaller minority party is transferred to a larger majority party. Not surprisingly, there are counterparts resolving a tie by taking a seat away from a larger minority party and allocating it with a smaller majority party.[19]

A first such procedure is the divisor method with geometric rounding (Hill), used in the USA since 1941 for the apportionment of the 435 seats in the House of Representatives to the 50 States. Applying this method to a delegation of size 16, the faction sizes 249 : 247 : 55 : 47 are mapped into 7 : 6 : 2 : 1 seats (divisor 38.3). Size 16 is the only tie situation resolved by this method, for the range considered inTable 1.[20]

A second method is the divisor method with 0.4-rounding (Condorcet), which also produces the final result 7 : 6 : 2 : 1 (divisor 38.8). This method resolves two of the fifteen ties listed in Table 1.[21]

A third procedure is the divisor method with rounding up (Adams), resolving five of the fifteen tie situations. The method is used in France to apportion the seats of the Assemblé Nationale to the Départments.[22]

There are committee sizes for which neither the divisor method with geometric rounding (Hill) nor the one with 0.4-rounding (Condorcet) resolves the tie. Moreover, it is possible that both methods do resolve a tie, but differently. An example is the German Bundestag 2002 at the beginning of the legislative period, with the then faction sizes 251 : 248 : 55 : 47. For a committee of size 36, the divisor method with standard rounding (Sainte-Laguë/Schepers) leads to the tie 15 : 15 : 3 : 3 (divisor 17). If we attempt to resolve the tie by using the two methods mentioned above, we get two conflicting answers: The divisor method with rounding down (D'Hondt) yields 16 : 15 : 3 : 2 (divisor 15.68), while the divisor method with rounding up (Adams) leads to 15 : 14 : 4 : 3 (divisor 17.8).[23]

---

clarification. But confronting the Court with the state-of-the-art raises the "danger", for the appellant, that the Court revokes not just a single D'Hondt apportionment, but the whole D'Hondt method.

[19]Marshall/Olkin/Pukelsheim (2002).

[20]Balinski/Young (2001, page 48). — *Joseph Adna Hill* (1860–1938) was Chief Statistician, Division of Revision and Results, US Bureau of the Census. — Sample calculation: The quotient $249/38.3 = 6.5$ lies above the decision point $\sqrt{6 \cdot 7} = 6.48$ and hence is rounded up to 7, while $247/38.3 = 6.45$ is rounded down to 6. The quotient $55/38.3 = 1.44$, when compared with the decision point $\sqrt{1 \cdot 2} = 1.41$, rounds up to 2, while $47/38.2 = 1.2$ goes down to 1. The decision points are *geometric means* of two neighboring integer numbers, whence the method receives its name.

[21]Balinski/Young (2001, page 63). — *Marie Jean Antoine Nicolas Caritat, Marquis de Condorcet* (1743–1794) was one of the leading politicians during the French Revolution. — Sample calculation: Fractions are rounded down when smaller than 0.4, and rounded up otherwise. Thus we get $249/38.8 = 6.42 \nearrow 7$ and $247/38.8 = 6.37 \searrow 6$ and $55/38.8 = 1.42 \nearrow 2$ and $47/38.8 = 1.2 \searrow 1$.

[22]See Balinski (2004a, page 190). — *John Quincy Adams* (1767–1848) was the sixth President of the USA.

[23]The gentle majority clause yields 16 : 14 : 3 : 3 (majority divisor 16, minority divisor 17.1).

As a consequence we refrain from a proposal to remedy the prevailing majority clause by taking recourse to a multitude of different apportionment methods. When many methods are tendered like on a flea market, many answers are conceivable: at best a unique and clear-cut tie break, or otherwise no tie break at all, or else multiple but conflicting results. A methodological zoo degenerates into a game of numbers. Instead the focus ought to be on electoral principles such as success-value equality, set forth by the German Federal Constitutional Court in 1952 and since then having generated an impressively consistent body of constitutional decisions.

8.   Minimum Seat Requirements

As a final point we would like to draw attention to the problem of guaranteeing each participant a minimum number of seats. With current faction sizes $249 : 247 : 55 : 47$ and for a committee of size 10, the divisor method with standard rounding (Sainte-Laguë/Schepers) results in the tie $4 : 4 : 1 : 1$ (divisor 60). The prevailing majority clause would resort to the divisor method with rounding down (D'Hondt), giving $5 : 4 : 1 : 0$ (divisor 49.6) and thus excluding the smallest party from representation.

However, the present problem concerns a committee of size 16, for which the Sainte-Laguë method yields the tie $7 : 7 : 1 : 1$. Considering how the divisor method with rounding down (D'Hondt) transfers seats from smaller to larger parties, there are just two possibilites: either the tie persists, or else it is broken into $8 : 7 : 1 : 0$. That is, the only way in which the prevailing majority clause could have resolved the tie would have deprived the smallest party of being represented at all. This may have set off some legal action of a different sort.[24]

It is easy to augment the gentle majority clause by the additional restriction that each participant be guaranteed representation. All that needs to be done is to modify the (unconditional) divisor method with standard rounding (Sainte-Laguë/Schepers), by demanding the minimum requirement that every participant receive at least one seat.[25]

We conclude with a *ceterum censeo*. The current topic, the apportionment of committee seats, is important. However, more important is the apportionment

---

[24]It is not clear to us how the Federal Constitutional Court would have settled the case. The Court sees the Conference Committee as a parliamentary body *sui generis*, for which the Constitution mandates neither a preservation of the majority (2 BvE 3/02, Rn. 67), nor a representation of all parliamentary groups, see BVerfGE 96 (1998) 264–288.

[25]The minimum committee size then is 5, of course, with the four parties filling one seat each and the fifth seat establishing a majority. The apportionments turn out to be $2 : 1 : 1 : 1$ for a committee of size 5, next $3 : 1 : 1 : 1$ for size 6, then $3 : 2 : 1 : 1$ for size 7, and finally $4 : 2 : 1 : 1$ for size 8. For committee sizes larger than 8 the apportionments of Table 1 apply.

of the Bundestag seats proper. The two-ballots electoral system of the German Federal Electoral Law is a top-quality product, enjoying high international esteem and serving as a prototype system.[26] But even top-quality products need be attended to. Negative weights of a ballot, doubly successful ballots, and overhang seats damage the image of the system.[27]

These deficiencies disappear when the idea of imposing minimum requirements is followed up. A simple adaptation of the divisor method with standard rounding would do, namely, imposing the minimum restrictions that each list receives at least as many seats as have been won in the constituencies. The *direct-seat restricted method* leaves no room for negative ballot weights, doubly successful ballots, nor overhang seats, and yet it stays in close harmony with the principle of success-value equality.[28] Whatever the requirements, the common denominator is the divisor method with standard rounding (Sainte-Laguë/Schepers). The method is so powerful that a few amendments suffice to adjust it to all practical purposes.

References

Balinski, M.L. (2004a): *Le Suffrage universel inachevé.* Paris.

Balinski, M.L. (2004b): "Die Mathematik der Gerechtigkeit." *Spektrum der Wissenschaft,* March 2004, 90–97.

Balinski, M.L./Young, H.P. (2001): *Fair Representation. Meeting the Ideal of One Man, One Vote. Second Edition.* Washington, DC.

Burkhardt, J. (1998): "Das größte Friedenswerk der Neuzeit. Der Westfälische Frieden in neuer Perspektive." *Geschichte in Wissenschaft und Unterricht,* 49, 592–612.

Fehndrich, M. (1999): "Paradoxien des Bundestags-Wahlsystems." *Spektrum der Wissenschaft,* February 1999, 70–73.

Fromme, F.K. (1970): "Regierungsmehrheit heißt nicht Ausschußmehrheit." *Frankfurter Allgemeine Zeitung,* Nr. 238 of 14 October 1970, 3.

Heckel, M. (1978): "*Itio in partes* — Zur Religionsverfassung des Heiligen Römischen Reiches Deutscher Nation." *Zeitschrift der Savigny-Stiftung für Rechtsgeschichte, Kanonistische Abteilung,* 64, 180–308.

Kämmerer, J.A. (2003): "Muss Mehrheit immer Mehrheit bleiben? Über die Kontroversen um die Besetzung des Vermittlungsausschusses." *Neue Juristische Wochenschrift,* 56, 1166–1168.

Lang, J. (2005): "Spiegelbildlichkeit versus Mehrheitsprinzip?" *Neue Juristische Wochenschrift,* 57, 189–191.

Lovens, S. (2003): "Die Besetzung der Bundestagsbank des Vermittlungsausschusses." *Zeitschrift für Parlamentsfragen,* 34, 33–41.

Mangoldt, H. von/Klein, F./Starck, C. (2005): *Das Bonner Grundgesetz: Kommentar. Fünfte Auflage.* München. Forthcoming.

---

[26] Shugart/Wattenberg (2001).

[27] Fehndrich (1999).

[28] Pukelsheim (2003, 2004a, b).

Marshall, A.W./Olkin I./Pukelsheim, F. (2002): "A majorization comparison of apportionment methods in proportional representation." *Social Choice and Welfare*, 19, 885–900.

Meyer, H. (2005): "Judex non calculat – Der Zweite Senat und die Besetzung des Vermittlungsausschusses." In: *Summa – Dieter Simon zum 70. Geburtstag.*, Editors R.M. Kiesow/R. Ogorek/S. Simitis, Frankfurt am Main, 405-433.

Pukelsheim, F. (2000a): "Mandatszuteilungen bei Verhältniswahlen: Erfolgswertgleichheit der Wählerstimmen." *Allgemeines Statistisches Archiv*, 84, 447–459.

Pukelsheim, F. (2000b): "Mandatszuteilungen bei Verhältniswahlen: Vertretungsgewichte der Mandate." *Kritische Vierteljahresschrift für Gesetzgebung und Rechtswissenschaft*, 83, 76–103.

Pukelsheim, F. (2000c): "Mandatszuteilungen bei Verhältniswahlen: Idealansprüche der Parteien." *Zeitschrift für Politik*, 47, 239–273.

Pukelsheim, F. (2002): "Die Väter der Mandatszuteilungsverfahren." *Spektrum der Wissenschaft*, September 2002, 83.

Pukelsheim, F. (2003): "Erfolgswertgleichheit der Wählerstimmen? Der schwierige Umgang mit einem hehren Ideal." *Stadtforschung und Statistik*, January 2003, 56–61.

Pukelsheim, F. (2004a): "Erfolgswertgleichheit der Wählerstimmen zwischen Anspruch und Wirklichkeit." *Die Öffentliche Verwaltung*, 57, 405–413.

Pukelsheim, F. (2004b): "Das Kohärenzprinzip, angewandt auf den Deutschen Bundestag." *Spektrum der Wissenschaft*, March 2004, 96.

Pukelsheim, F./Maier, S. (2005): "Eine schonende Mehrheitsklausel für die Zuteilung von Ausschusssitzen." *Zeitschrift für Parlamentsfragen*, 36, 763-772.

Riescher, G./Ruß, S./Haas, M.A. (2000): *Zweite Kammern*. München.

Schindler, P. (1999): *Datenhandbuch zur Geschichte des Deutschen Bundestags 1949 bis 1999*. Baden-Baden.

Schuster, K./Pukelsheim, F./Drton, M./Draper, N.R. (2003): "Seat biases of apportionment methods for proportional representation." *Electoral Studies*, 22, 651–676.

Shugart, M.S./Wattenberg, M.P. (2001): *Mixed-Member Electoral Systems: The Best of Both Worlds?* Oxford.

Stein, K. (2003): "Die Besetzung der Sitze des Bundestages im Vermittlungsausschuss." *Neue Zeitschrift für Verwaltungsrecht*, 22, 557–562.

Allotment According to Preferential Vote: Ecuador's Elections

Victoriano Ramírez

Departamento de Matematica Aplicada, University of Granada

Abstract   In this paper we show results for some unipersonal and pluripersonal elections in Ecuador and in Spain, observing that the used methods can be replaced by better ones. We present a method for individual elections based on one-on-one comparisons, and preferential voting. This method verifies CONDORCET and PARETO, and it is furthermore better than the Two Round method. For multicandidate elections we give proportional and monotone methods based on preferential vote. We also analyse the electoral system of Ecuador and propose alternatives for it.

Keywords: Electoral Systems, Preferential Vote, Borda, Condorcet, Sample Transferable Vote, Proportionality.

1.   Introduction

Results from some unipersonal and pluripersonal elections in Ecuador, and similar results in others countries such as France, Spain, etc., show that the currently used methods cannot be satisfactory.

In unipersonal elections, the Two Round method is frequently used, and normally the two candidates participating in the second round are the most preferred by the voters; but this is not always true. The second round between Chirac and Le Pen in the 2002 Presidential in France is a well known case: the three most voted candidates obtained a very similar percentage of votes, concretely J. Chirac obtained 19.71 %, J. M. Le Pen 16.95 % and L. Jospin 16.12 %.

There are many unipersonal elections in which three or more candidates obtain a close percentage of votes (as in the 2002 Presidential election in Ecuador), and so we can not affirm in general that the most preferred candidate is elected when we use the Two Round method (or TR method).

In this paper we show that there exists a method that is better than the TR method for unipersonal elections. It is a method based on one-on-one comparison according to an agenda and therefore the method verifies CONDORCET.

On the other hand, in multicandidate elections, Approval Voting (AV) (Brams and Fishburn, 1983) or Limited Approval Voting methods (LAV) are sometimes used (in LAV the number of candidates that the voters can vote for is limited). For example, LAV is used for political elections in Ecuador, for the Senate election in Spain, for university elections in Spain, etc. In such cases low proportionality is obtained; the Senate election in Spain is an example (Ramirez and Palomares, 2006). When proportionality is required, the Single Transferable Vote (or STV) can be used, but STV is not monotone; that is, a candidate A can obtain a seat when n representatives must be elected, but A does not win when the number of representatives is $n + 1$.

In this paper, for unipersonal election we establish some desirable properties such as PARETO, CONDORCET (Taylor A., 1995, pg. 106), etc..., and we describe a procedure based on pairwise comparison of candidates according to an established agenda. This method is better than the TR method in a given sense.

For multicandidate elections we show certain voting procedures with a proportionality equivalent to some classical proportional methods such as Webster and Jefferson. Then, manipulation via cyclical voting and the thresholds of representation are analyzed. We give proportional methods based on preferential vote, whose threshold of representation is less than that corresponding to the methods AV and LAV.

In addition, we analyse the elections for the President and the Congress in Ecuador and we offer alternatives.

2.   Unipersonal Election

2.1   The 2002 Presidential Election in Ecuador

In the 2002 Presidential election in Ecuador, six candidates obtained voting percentages between 11.9% and 20.6 % in the first round. The exact results were (http://www.tse.gov.ec, http://www.electionworld/ecuador.htm) (Table 1).

The two first candidates in the first round were Gutierrez and Noboa; next, in the second round Gutierrez beat Noboa with 54.8% of the votes. Edwing Gutierrez was the President of Ecuador from 2002 to 2005. We put forth some questions:

o Would Gutierrez have beaten Roldos, Borja or Neira, etc., in a one-on-one competition?

o Was E. Gutierrez the most desired candidate in the election of 2002?

Table 1. Results for the 2002 presidential election in Ecuador.

Candidate	First Round	Second Round
Lucio Edwing Gutierrez Barbua	20.6%	54.8 % Winner Gutierrez
Álvaro Fernando Noboa Ponton	17.4%	45.2%
Leon Roldos Aguilera	15.4%	
Rodrigo Borja Cevallos	14.0%	belongs to ID (3th party in the Congress)
Antonio Xavier Neira Menendez	12.1%	belongs to PSC (1st party in the Congress)
Jacobo Bucaram Ortiz	11.9%	belongs to PRE (2nd party in the Congress)
Five Remaining candidates	8.6 %	
	100%	

- Possibly: Yes.
- Possibly: No (we do not know the answer)

In the case at hand, E. Gutierrez belongs to a political party with few deputies in the Congress. The Congress terminated Gutierrez's employment on April 20, 2005.

If the TR method can not guarantee satisfactory answers, we may consider the STV method. But is the STV method consistently better than the TR method? In what way?

Verifiying the Condorcet criterion: **No.**

EXAMPLE 1

Preferences	Votes
$A > B > C > D$	35
$B > D > A > C$	30
$C > B > D > A$	25
$D > C > B > A$	10

Here B is the Condorcet winner, being the winner under the TR Method.

Under the STV method, D is the first candidate to be eliminated, followed by B, meaning that A is the winner (because A triumphs over C)

2.2   Properties for Unipersonal Elections

We say that a method *M is better than the TR method* (in unipersonal elections) if, for each election:

M gives the same solution as that of the TR method

or

the winner under method M is more desired than each one of the candidates taking part in the second round.

This means that if A and B are admitted to the second round, and C is the winner with method M, then:

○ C is equal to A, or

○ C is equal to B, or

○   C wins over A (in a one-on-one competition), and

    C wins over B (in a one-on-one competition).

But this is not the only property that we wish to guarantee. We are looking for a method M that:

a. Verifies CONDORCET,

b. Verifies PARETO (unanimity) and moreover

c. Must be better than the TR method, in the previously established sense.

3.   The One-on-One Comparison Method

One possible solution to the above problem is very simple: *Comparing the candidates one against one according to an agenda (or an order).*
On the basis of the first preference we make an agenda for the comparisons (if they are equal then the second preference is used . . .). More precisely, for each candidate $i = 1, \ldots, m$, we define the vector $v_i = (v_{i1}, v_{i2}, \ldots, v_{im})$, where v_{ik} is the number of votes (possibly 0) that rank i as the k-th candidate, and then we stipulate that candidate i precedes candidate j in the agenda if v_i is lex-greater than v_j. So, for Example 1 the agenda would be: ABCD.
The comparisons are performed backwards in the agenda.
Therefore, for Example 1, the method begins by comparing C with D (in this case, C wins over D, 60-40); then it compares B with C (here 65-35); and goes on to compare A with B (here 35-65). Thus B is the winner with this method.
When in the last comparison of the agenda, the first candidate of the agenda wins (A in the previous example) and we have not made the comparison between the first two candidates (A and B, above) we must look at an additional comparison. For example:

EXAMPLE 2

Preferences	Votes
A > B	60
B > D > C > A	40
C > A > D > B	50

(The preference A>B is equivalent to saying $A > B > C = D$).

Now, the agenda is : ACBD

Comparison BD: 100-50, winner B.

Comparison CB: 50-100, winner B.

Comparison AB: 110-40, winner A. The additional comparison AC is needed.

Additional comparison AC: 60-90, winner C. So C is the winner with this method (in this case C is also the winner under the TR Method).

EXAMPLE 3 *In the following case:*

Preferences	Votes
A > B	60
B > D > C > A	40
C > D > B	50

the agenda is ACBD, and the winner after the first three comparisons is B. Therefore, no additional comparison is needed. In this example, B is the Condorcet winner.

EXAMPLE 4 *Finally, we consider the case:*

Preferences	Votes
A	24
B	22
C	20
D > E	12
D > C	6
E	16

The agenda is ABCDE, and the winner after four comparisons is C. In this example, the Condorcet paradox can be seen.

There are agendas for which PARETO (unanimity) is not verified (Saari, 1994 and Taylor, 1995). Trivially, with the proposed agenda, *PARETO is always verified.*

Now, the following result is obvious:

THEOREM 5 *The method based on one-on-one comparison according to the previous agenda verifies the properties established above (that is, a., b. and c. of section 2.).*

4.   Choosing Several Representatives. Results for Different Elections

We will show some results from multi-candidate elections, in Spain and in Ecuador, and look for new properties, such as proportionality.

4.1   The Elections in the Spanish Universities

How does the electoral method work in Spanish universities?
It is very similar in most Spanish universities.

○ All the candidates appear on the same list.

○ The voters must vote for a maximum number of candidates. For example,

  – To choose 50 representatives, the voters are able to vote for a maximum of 38 candidates (this figure is likely to change, depending on the different universities)

  – There are usually between 150 and 400 voters.

○ However, normally less than 40% of the electors vote.

○ The candidates with the most votes are chosen.

Manipulation. A very beneficial strategy: voting cyclically
Sometimes a group of candidates may be organised by a person that knows, very well, how to manipulate the electoral method. For example:

> In a Center of Granada University, in 1998, a Teacher chose 50 candidates, let us say from $c_1, \ldots$, to c_{50}. He explained to them how to vote cyclically. In this way, c_1 must vote from c_1 to c_{38} (each one of them obtains a point for the vote of c_1); c_2 must vote from c_2 to $c_{39}, \ldots$; similarly, c_{20} must vote from c_{20} to c_{50} and from c_1 to c_7.

- Therefore, the votes of the 50 chosen candidates result in 38 points to each one of them.

- Meanwhile, each one of the remaining candidates usually votes for him or herself, and he or she obtains one point.

- The rest of the electors are not organized and they can not modify the result. This means that the 50 candidates who voted cyclically would be the winners.

- In this Center precisely 50 (organized) candidates will obtain representation, gaining between 46 and 65 votes each (many of the new representatives would be unknown due to the fact that they are new teachers).

- The other candidates would obtain less than 19 votes.

Numerically and graphically the results of this election in 1998 are shown in Table 2 and Figure 1.

Table 2.   Example of manipulated university election in Spain using AVL method.

Votes	0 - 9	10 - 18	19 - 27	28 – 36	37 - 45	46 - 54	55 – 65
Candidates	8	10	0	0	0	**33**	**17**
Elected?	no	no	-	-	-	**yes**	**yes**

This leads us to the following question:

If there are 350 voters, for example, how many of them must vote only for c_{51} to ensure him/her a representative' position when the remaining voters vote cyclically? It is very easy to see that over 150 are needed (more than 43% of them), a very high threshold of representation.

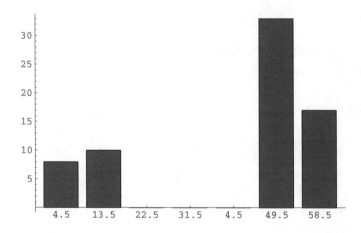

Fig. 1.   Graph corresponding to the example of Table 2 (manipulated election).

Few changes and little proportionality.

The previous example is very special as many of these candidates were new teachers (organized by veteran teachers). Sometimes a very important aspect of the organized candidates is that they were representatives during the previous period.

For this reason, in Spanish universities many people continue to be representatives time and time again.

When there is only one organised group, it can obtain almost all the representative positions. If there are two or more organised groups, the major group with 50 or more candidates and the other groups with less than 38 (for example 20 or less per each), it is possible for the first group to obtain all positions. In any case: a great margin for manipulation and no proportionality at all.

4.2   The Election of Ecuador's Congress

For the election of the Ecuadorian Congress

- There is a ballot containing all of the candidates from all of the political parties.

- Each voter can vote for candidates belonging to one party or belonging to several political parties.

- Each elector can vote for a number of candidates equal to the size of their constituency.

- Usually, each voter will vote for candidates that belong to the same political party. Examples:

The size of the Congress in Ecuador is 100 and the number of constituencies is 22.

The name and the size of the five greatest constituencies can be seen in Table 3 :

Table 3.   The five greatest constituencies in Ecuador.

Constituency	Guayas	Pichincha	Manabi	Los Rios	Azuay	Other
Deputies	18	14	8	5	5	Less than 5

Some results from the election of 2002 were (The Supreme Electoral Court in Ecuador, http://www.tse.gov.ec and http://www.electionworld/ecuador.htm):

Guayas: 18 deputies
The 18 candidates with the most votes were (all of them belonging to PSC):

1st Febres	**PSC**	10th Ramirez	**PSC**
2nd Harb	**PSC**	11th Ordonez	**PSC**
3rd Cioppo	**PSC**	12th Martillo	**PSC**
4th Viteri	**PSC**	13th Gamboa	**PSC**
5th Jaramillo	**PSC**	14th Concha	**PSC**
6th Davila	**PSC**	15th Jaya	**PSC**
7th Varas	**PSC**	16th Salazar	**PSC**
8th Moran	**PSC**	17th Mussuh	**PSC**
9th Valverde	**PSC**	18th Samaniego	**PSC**

The most voted, Febres, obtained 461,316 votes, ..., the 18th, Samaniego, 348,273 votes; the 19th was Carrera belonging to PRIAN who gained 221,815 votes. For the Andean Parliament election (party list), which took place simultaneously, the results for PSC and PRIAN were: PSC = 422,179 votes, PRIAN = 188,898 votes.

In many other constituencies all of the most voted candidates belonged to the same political party, shown in Table 4.

Table 4. Parties of the most voted candidates.

Constituency	Size (h)	Party of the h most voted candidates
Pichincha	14	ID ID ID ID ID ID ID ID ID ID ID ID ID ID
Manabi	8	PSC PSC PRE PSC PSC DP-UDC PRE PRE
Los Rios	5	PRE PRE PRE PRE PRE
Azuay	5	ID ID ID ID ID

Only in Manabi did the h most voted candidates belong to two or more political parties.

If we assign the seats of deputy to the h most voted candidates, it may be that only one political party obtains all of the representatives - a total lack of proportionality.

Nevertheless, Ecuador's Constitution states in its article 99 that the number of deputies for each political party must be proportional. By virtue of this, in Ecuador proportional apportionment is applied to the sum of the votes of the candidates of each political party. Hence, these methods have been used:

- Before 2004, Ecuador applied Jefferson's method (in February 2004, Jefferson's method was declared inconstitutional).

- After 2004 (October), a method was introduced with the divisors: 2, 3, 4, ... proposed by the TSE.

In the above examples the voters can vote for a number of candidates equal to or close to the number of representatives to be chosen (the LAV method). We underline that LAV can prove very manipulable (as in the Spanish universities), and not proportional when the voters vote as in elections based on lists of political parties (as in the Spanish Senate (Ramirez and Palomares, to appear) and the Ecuadorian Congress).

5. Proportional Methods Based on Preferential Votes

5.1 Accepting the Proposed Order, (AO)

- Let us suppose that the candidates form groups (or political parties) g_1, $g_2, \ldots, g_n$, not necessarily of the same size.

- We say that the voters behave by **accepting the proposed order (AO)** if each voter chooses only candidates in one group (or political party) and they accept the order in which the candidate names appear (Ramirez and Palomares, 2006). For example, if the groups and the candidates are the following ones:

Groups (or Lists)	Candidates
g_1	$c_{11}c_{12}\ldots c_{1r}$
g_2	$c_{21}c_{22}\ldots c_{2s}$
$\ldots$	$\ldots$
g_n	$c_{n1}c_{n2}\ldots c_{nt}$

The electors have an AO behaviour if each voter casts a partial preferential vote as $c_{11} > c_{12} > \ldots > c_{1r}$, or $c_{21} > c_{22} > \ldots > c_{2s}, \ldots$, or $c_{n1} > c_{n2} > \ldots > c_{nt}$

That is, if the results are as follows:

EXAMPLE 6

Preferences	Votes
$c_{11} > c_{12} > \ldots > c_{1r}$	v_1
$c_{21} > c_{22} > \ldots > c_{2s}$	v_2
$\ldots$	
$c_{n1} > c_{n2} > \ldots > c_{nt}$	v_n

5.2 AO and Proportional Social Choice for Preferential Votes

- Let us assume a social choice method S based on preferential vote and a proportional apportionment method P. Let us suppose that the electors have an AO behaviour (as before).

- Then we say that the social choice method S *has the same proportionality* as the proportional method P if the result using S (for the preferences) is the same as if we use the P method (for the votes $v_1, v_2,..., v_n$). Therefore, we say that the method S is P -Proportional.

Answer using S (for preferences) = Answer using P for the votes ($v_1, v_2,...,$ v_n).

EXAMPLE 7

In order to choose four representatives, let us suppose that 100 electors vote for nine candidates in this way:

Preferences	Votes
$c_1 > c_2 > c_3 > c_4$	50
$c_5 > c_6 > c_7$	35
$c_8 > c_9$	15

The quotas for the three groups are: q = (2.0, 1.4, 0.6).
Then,

- If S has the same proportionality as Webster (or Saint-Laguë), the answer must be: c_1, c_2, c_5 and c_8 (when we apply S to the preferences).

- If S has the same proportionality as Jefferson (or d'Hondt) the answer must be: c_1, c_2, c_5 and c_6.

The Approval Voting method is not proportional (the answer, in this case will be c_1, c_2, c_3 and c_4). Likewise, LAV is not proportional.

5.3   The Proportional Single Transferable Vote (STV)

STV has the same proportionality as that of:

- The Droop method, when it uses the Droop quota.

- The Hamilton method (or Greatest Remainder method), when it uses the Hare quota.

5.4   Borda-Type Methods for Proportional Choice

- A Borda-type method uses a sequence of positive numbers (the weights): $w_1 > w_2 > ... > w_n$

- If an elector votes A>B>C..., then

  – candidate A adds w_1 points

– candidate B adds w_2 points, etc.

THEOREM 8

We have obtained the following results (Ramirez and Palomares, to appear):

o The Borda-Type method with weights: $w_1 = 1$, $w_2 = \frac{1}{3}, \ldots, w_i = \frac{1}{2i-1}, \ldots$ (or Borda-Webster methods) has the same proportionality as Webster's method.

o The Borda-type method with weights: $w_1 = 1$, $w_2 = \frac{1}{2}, \ldots, w_i = \frac{1}{i+1}, \ldots$ has the same proportionality as Jefferson's method.

o Similarly for other divisor methods.

If we apply the Borda-Webster and Borda-Jefferson methods to the Example 6, we obtain the points that are shown in Table 5, in bold the elected candidates.

Table 5. Points obtained by the candidates from Example 6.

Candidate	c_1	c_2	c_3	c_4	c_5	c_6	c_7	c_8	c_9
Borda-W	**50**	**16.7**	10	7.1	**35**	11.7	7	**15**	5
Borda-Jef	**50**	**25**	16.7	12.5	**35**	**17.5**	11.7	15	7.5

The answers are the same when we apply, respectively, Webster's method and Jefferson's method (Balinski and Young, 1982) to assign four seats in proportion to the votes (50, 35, 15).

6.   Threshold of Representation for Different Methods

Now, let us suppose that four representatives must be chosen. Then, one question is:

How many voters must vote for c_5 (only) to assure a win to c_5, if the rest of the voters vote cyclically for c_1, c_2, c_3 and c_4?

If we use the Borda-Webster method, and there are 100 voters, the best strategy for c_1, c_2, c_3 and c_4 would be equal points for all of them, that is:

$$c_1 > c_2 > c_3 > c_4 \qquad k \text{ votes}$$
$$c_2 > c_3 > c_4 > c_1 \qquad k \text{ votes}$$
$$c_3 > c_4 > c_1 > c_2 \qquad k \text{ votes}$$
$$c_4 > c_1 > c_2 > c_3 \qquad k \text{ votes}$$
$$c_5 \qquad\qquad\qquad 100 - 4k \text{ votes}$$

Then, c_1, c_2, c_3 and c_4 obtain $k(1 + \frac{1}{3} + \frac{1}{5} + \frac{1}{7})$ points each, and c_5 obtains $100 - 4k$ points. Therefore, k must be greater than 17. So, 29 votes for c_5 would always ensure a win to c_5.

Next, in Table 6, the threshold of representation will be shown with regard to Borda-Webster method, Borda- Jefferson method, Limited Approval Voting (limited to 70% of the constituency's size) and Approval Voting methods.

Table 6. Threshold of representation as percentages for different methods.

Size	Borda-Webs.	Borda-Jef.	LAV-70%	AV
10	17.6%	22.7%	41%	50%
20	11.0%	15.2%	41%	50%
40	6.6%	9.7%	41%	50%
80	3.8%	5.8%	41%	50%

7. Some Remarks Regarding Borda-Type Proportional Methods

o A unique Borda-Type proportional method exists and can be applied to a preferential vote with the same proportionality as each divisor method in proportional representation. The weights are the inverse of the divisors.

o No Borda-Type method verifies CONDORCET.

o All Borda-Type methods are monotone with respect to the number of representatives.

o Borda-Type methods have a threshold of representation lower than those of AV and LAV.

8. Other Proportional Methods Based on Preferential Vote

We can obtain proportional methods based on preferential votes (total or partial) as a generalization of the one-on-one comparison method proposed for unipersonal elections. Firstly we establish the agenda for the comparisons. The representatives are obtained one by one. When a candidate gets a seat, he or she is eliminated from the agenda.

In this case we must use weights (as in Borda-Type methods) in the comparisons to obtain the second, the third, etc. representatives. The weights are applied according to the previous representatives elected.

Therefore, in practice, a computer is required for the application of this method.

Example 7. Selecting two candidates according to the following preferences

Preferences	Votes
A > B >	10
D > C	10
B > A > C	10
C > A > B	9.

Then, if we use the Borda-Webster method the answer is: A and B. If we use the one-on-one comparison method, the agenda is ABDC and the first selected candidate is A; so the new agenda is BDC and the second winner (using one-on-one comparison with the same proportionality as Webster) is C.

9. Ecuador's Electoral System for Congress: a Brief Analysis

The sizes of the 22 Constituencies for Ecuador's Congress are in Table 7:

Table 7. The sizes of the 22 constituencies in Ecuador.

Guayas	18	Chimborazo	4	Imbabura	3	Pastaza	2
Pichincha	14	Cotopaxi	4	Carchi	3	Zamora	2
Manabi	8	Tunguragua	4	Bolivar	3	Marona-Sant.	2
Los Rios	5	El Oro	4	Orellana	2	Napo	2
Azuay	5	Loja	4	Sucumb.	2	Galapagos	2
Esmeraldas	4	Cañar	3				

And the results of the 2002 Congressional election are shown in Table 8.

9.1   Remarks on Ecuador's Electoral System

○ In the last presidential election of 2002, six candidates obtained a close percentage of votes in the first round. Surely, in this case (also in other similar cases) the one-on-one comparison method gives a better result than that of the TR method.

○ For the Congress,

– Ecuador's Constitution favours proportionality and makes it possible to vote for candidates that belong to different political parties.

Table 8.   The 2002 Congress Election in Ecuador.

Social-Christian Party,	PSC	24
Ecuatorian Roldosist Party	PRE	15
Party of the Democratic Left	ID	13
Institutional Renewal Party of National Action	PRIAN	10
Coalition	MUPP-NP/PSP	6
Pluri-National Pachakutik United Movement-New Country,	MUPP-NP	5
People's Democracy-Christian. Union-Popular Democracy,	DP-UDC	4
Democratic People's Movement	MPD	3
Patriotic Society Party	PSP	2
Joint list of MUPP-NP and PS-FA		2
Joint List of MPD and PS-FA		2
Social Party-Broad Front	PS-FA	1
Concentration of People's Forces	CFP	1
Provincial Integration Movement	MIP	1
Solidary Fatherland Movement	MPS	1
Democratic Transformation	TD	1
Joint list of ID and MIRE		1
Joint list of ID and DP-UDC		1
Joint list of ID and NP		1
Joint list of MUPP-NP and MCNP		1
Joint list of DP-UDC, PS-FA and AN		1
Joint list of PSC and UN-UNO		1
Joint list of PSC and AN		1
Joint list of PSP and MPD		1
		100

- 60% of the deputies were chosen in constituencies whose size is less than or equal to 5. Therefore, in these cases the proportionality in each constituency is low.

- Usually each elector votes for candidates of the same political party. Therefore, if we just applied AV or LAV (limited to the constituency size), then the largest party would obtain all of the representatives in the constituency.

10.   Conclusions

○ In unipersonal elections, when we use the comparative one-on-one method, CONDORCET is verified; and this method does not harm in a significant way the presence of candidates that compete for the same votes.

○ In pluripersonal elections, the Borda-Webster method is proportional and monotone. To obtain proportionality, the Borda-Webster method may be preferred when the size of the constituency is large.

○ However, when there are many small constituencies, as for the Ecuadorian Congress, the proportionality is low and other techniques can prove more adequate. For example, for Ecuador's Congress we might:

- Not change the Constitution.

- Not change the form of voting. That is, use the current LAV method, or the Approval Voting one proposed by S. Brams (always with a normalisation of votes). This is a *very easy method for the voters*.

- Establish a threshold (fewer parties in the Congress). Beyond this all votes must count equally.

- Change (in the "Ley de Elecciones" = electoral law) the formula for assigning the seats and apply *complex formulas to obtain fair results*, if necessary. For example, apply the Bi-Proportional apportionment (Balinski and Demange, 1989) to the sum of the points (normalised votes) of the total candidates of each political party. Usually, the BAZI algorithm developed by F. Pukelsheim (www.math.uni-augsburg.de/stochastik/bazi/pseudoCode.html), among others, finds the number of representatives of each political party in each constituency that must be assigned. Finally, the seats of a party are assigned to the candidates who receive the most votes.

Acknowledgments

The author wishes to thank the Junta de Andalucia (regional government) for support related to research on proportional representation and social election, through the FQM-191 group. The author also thanks the anonymous referee for helpful suggestions which improved the paper.

References

Balinski, M. L., Young H. P. (1982). *Fair Representation: Meeting the Ideal of One Man One Vote*. New Haven, CT.

Balinski, M. L., Demange, G. (1989). "Algorithms for proportional matrices in real and integers" *Mathematical Programming*, 45, pp. 193-210.

Brams, S. & Fishburn, P. (1983). *Approval Voting*, Birkhäuser, Boston.

Ramirez, V., Palomares, A.. The Senatorial election in Spain. Proportional Borda Methods for selecting several candidates, to appear in *Annals of Operation Research*.

Saari, D. (1994). *Geometry of Voting,* Springer-Verlag.

Taylor A. (1995). *Mathematics and Politics*, Springer-Verlag.

Degressively Proportional Methods for the Allotment of the European Parliament Seats Amongst the EU Member States

Victoriano Ramírez, Antonio Palomares, Maria L. Márquez
Departamento de Matematica Aplicada, University of Granada (Spain)

Abstract   In this work we present several methods for distributing the seats of the European Parliament amongst the States of the European Union, in accordance with the restrictions established in article I-20 of the projected European Constitution. The proposed methods can be applied to the current composition of the EU, but also if there is a change in the number of states or their populations. They are based on *adjusting the quotas* of every country so that they verify the constitutional restrictions, and so that their rounding to an integer number will constitute an allotment of the seats of the Parliament.

Keywords: European Parliament, European Constitution, Proportional Representation, Degressive Proportionality, Quota Adjustment, Webster, Jefferson.

1.   Introduction

The European Parliament is the institution that most resembles the Congress of any democratic country (but with important differences). The European Parliament exercises, jointly with the European Council, legislative and budgetary functions, as well as certain control functions, and it is in the charge of electing the President of the Commission. It has been growing in size with the successive enlargements of the EU, though the projected Constitution limits its size in the future to a maximum of 750 members.

The number of seats in the European Parliament must be distributed amongst the States of the EU in accordance with their populations. Traditionally, the distribution of the seats of the European Parliament does not follow proportionality criteria, as is usual when the seats of the Congress of a country are allotted amongst their constituencies; yet in the EU the small countries have been over-represented with respect to the large ones. For example, Luxembourg at present has six seats (as in previous legislatures), although its exact quota is less than one. On the other hand, countries with important differences in population have received the same number of representatives. Thus, France, the United Kingdom and Italy have the same number of representatives, although their populations are appreciably different (before its unification, the Federal Republic of Germany also used to receive the same number of seats as these countries).

How has the allocation of seats to the States of the European Parliament been established in every legislative period? By negotiation.

Obviously, negotiation is not the definitive solution. In fact, the Project of European Constitution itself states that the European Council must adopt a decision that fixes the composition of the European Parliament. Such a decision would, logically, contain a formula to determine the distribution of the seats of the member States. Moreover, it must be valid for the allotment of seats to the present States, while remaining valid if the number of UE countries, or the size of their populations, changes.

This formula would not be common to other problems of constituency allotment (http://www.publications.parliament.uk/pa/ld200203/ldselect/ldeucom/169/16921.htm), as the European Constitution has established criteria that impose *disproportionality* – perhaps, a disproportionality somewhat similar to that which results from the negotiations for previous and present allotments.

The aim of this work is to analyse with precision projected EU constitutional restrictions and, accordingly, to propose different methods for allotting the seats of the European Parliament to the EU States.

Firstly, in section 2 we analyse the Project of Constitution and then establish and introduce the concept of quota adjustment for a degressively proportional allotment. Next, we justify the established criteria. We tackle the problem of the "degressive proportionality" as that of obtaining a quota adjustment such that, when rounded using a proportional method, it gives rise to an allotment in accordance with the Constitution. That is, the degressive proportionality is reached by adjusting quotas. For rounding, we use the Webster method of proportional allotment, because it is the only method that is consistent and impartial, and the Jefferson method because it benefits the biggest countries (Balinski and Young, 1982). The Webster method – noted in the following by W – rounds every fraction to the closest whole number; the Jefferson method – noted in the following by J – rounds every fraction to its integer part.

In section 3, we proof that several adjustments of quotas (linear, parabolic, potential, etc.) allow degressively proportional allotment to be obtained according to the Project of Constitution. We show some numerical and graphical examples for the case of the 25 member States and some enlargements of the EU.

In section 4, we establish additional remarks to help in choosing a method to allot the seats of the European Parliament. Four possibilities have been developed in the previous sections: the parabolic and potential adjustment rounding with either W or J. All four are equally easy to apply in practice, though we believe that the parabolic one more closely resembles with the last negotiated allotment and we recommend it if we prefer not to give the advantage to the smallest countries.

2.   Restrictions and Temporary Regulations in the Project of European Constitution

The projected European Constitution[1] of October 2004 states, in its article I-20, that the European Parliament will be made up of Union citizens, and that the number of parliamentary members will not exceed 750. It also says that representation will be "degressively proportional", and every member State must receive a minimum of six deputies and a maximum of 96.

The Constitution has also foreseen the representation of Romania and/or Bulgaria if they become members of the Union before the European Council approves the composition of the European Parliament. In such a case, Romania will be represented by 35 Euro-deputies and Bulgaria by 18, so that the parliament will have 785 until the end of the 2004-2009 legislature.

The term *"degressively proportional"* is a recent concept that does not respond to a concrete formula of seat allotment (http://www.taemag.com/docLib/20040128_p4043.pdf), but to a limitation according to which the States with fewer inhabitants receive fewer representatives than the States with more inhabitants; yet the States with fewer inhabitants receive more representatives than they proportionally deserve, and the more populated States receive fewer representatives than they proportionally deserve (Bovens, L. "Welfare, Voting and the Constitution of a Federal Assembly", http://www.uni-konstanz.de/ppm/EU.pdf).

2.1   Notations and Definitions

Given the populations of the n States $P = (p_1, p_2, \ldots, p_n)$ and the size H of the European Parliament, the exact quota, x_i, corresponding to a country with population p_i is $x_i = \dfrac{p_i H}{T}$, T being the total number of inhabitants of the EU. Let m and M be the quotas respectively corresponding to the smallest and the biggest two countries. So, at the current EU, $m = 0.64$ corresponds to Malta, and $M = 132.92$ to Germany (see Table 1).

Definition 1: A *quota adjustment function* is a function $A : [m, M] \longrightarrow R$ transforming quotas x into other adjusted quotas $A(x)$.

---

[1]Tratado por el que se establece una Constitución para Europa, Ministerio de Asuntos Exteriores y de Cooperación, Ministerio del Interior y Ministerio de la Presidencia, Dep. Legal M-53614/2004.

Definition 2: In this paper, an allotment will be called degressively proportional if it is the result of applying a rounding proportional method using a non-decreasing concave quota adjustment function A.

2.2 Criteria for the Quota Adjustment Function A Derived from the Projected Constitution.

1 Increase

$$x < y \text{ implies that } A(x) < A(y), \ \forall x, y \in [m, M].$$

That is, if the exact quotas of two States are x and y, verifying $x < y$, then the corresponding adjusted quotas must verify the same relation.

2 Bounds

Meanwhile, if the exact quotas of all the countries belong to an interval $[m, M]$, it must be verified that:

$$6 \le [A(x)]_r \le 96 \quad \text{for all } x \text{ in the interval} \quad [m, M],$$

where $[A(x)]_r$ is the rounding of $A(x)$ with a proportional method.

We can therefore guarantee the maximum and minimum limits of seats that a country can receive.

3 Concavity

The quotas mean the exact proportionality, that is, the adjustment function $A(x) = x$ (whose plot is the bisectrix of the first quadrant, a line of slope *one*), whose function corresponds to the exact proportionality. The total absence of proportionality would be to assign the same number of representatives to every country, this corresponding to the function $A(x) = c$ (a constant function, so, with *zero* slope).

In this third sense, then, it is possible to interpret that the concept "degressively proportional" implies using a function $A(x)$ whose slope is decreasing, or at least not increasing. So, in the case that we use an adjustment function that can be twice differentiated, we require that

$$A''(x) \le 0, \quad \text{for all } x \in [m, M].$$

If $A(x)$ is piecewise rectilinear, the successive slopes of the polygonal would be smaller and smaller.

4 Assign H seats

$$\sum_{i=1}^{n} [A(x_i)]_r = H$$

That is, at least one of the roundings of $[A(x_i)]_r$ must sum H.

As, we will use the Webster or the Jefferson method, for the rounding. We will change the subscript r to W or J respectively.

Obviously, there are infinite possibilities for choosing the functions that *adjust the quotas* verifying the previous requirements (although the number of different allotments that they produce is finite). We try to choose the simplest possible adjustment functions.

3.   Several Quota Adjustments for Degressively Proportional Allotment in the EU

3.1   The Rectilinear Adjustment of Quotas

The simplest adjustment is the rectilinear one, that is, the adjustment that uses a function like $A(x) = a + bx$ (whose plot is a straight line). If, from the current populations of the 25 member States of the EU and $H = 732$, we consider the rectilinear adjustment that gives 96 seats to the largest country, we obtain that possible functions – rounding with Webster or Jefferson, A_W or A_J – verifying the four previous criteria, are:

$$A_W(x) = 96 + 0.644(x - 132.92) = 10.3995 + 0.644x,$$

$$A_J(x) = 96 + 0.64(x - 132.92) = 10.9376 + 0.64x,$$

In this case, the rounding of $A_W(x_i)$ with the Webster method and the rounding of $A_J(x_i)$ with the Jefferson method give the same results. Specifically, the allotment for the 25 countries will be:

11-11-11-12-12-13-14-15-16-16-16-19-20-21-21-21-21-22-27-50-53-
70-72-72-96

It is possible to make other rectilinear adjustments that sum up to 732, which is the actual size of the European Parliament, but such adjustments allocate 11 or more seats to the least populated countries. The next example shows one such adjustment:

$A(x) = 93 + 0.615(x - 132.92)$ produces the following rounding with W:

12-12-12-13-13-14-15-15-16-17-17-19-20-21-21-22-22-22-27-49-52-
68-70-70-93

With the first of these two adjustments, Germany receives 96 seats, while
the smaller States receive at least 11 seats, although their exact quota is
less than 6 (even smaller than 1 for some of them, as we can see in Table
1), and the projected Constitution only forces the assignment of six.

The second adjustment produces an allotment that is even less proportional with the current data. However, both adjustments verify the four
properties previously established – increasing, bounding, concavity and
sum equal to H – and so, it seems reasonable to find new adjustment
functions, in order to avoid giving more representation to very small
States, at least with the present situation.

It is easy to see that with the current population data, the allotments
that assign only 6 seats to the smallest country, using Webster for the
rounding, require an H value under 650. Specifically, for $H = 649$, we
can use

$$A(x) = 6.49 + \frac{90}{132.92 - 0.64}(x - 0.64) \simeq 6.05 + 0.68x$$

and we obtain

6-7-7-8-8-9-10-10-12-12-12-15-16-17-17-17-17-18-24-48-52-69-71-
71-96

Therefore, forcing *a rectilinear adjustment*, for $H \geq 650$, with the current populations of the States, and allotting only 6 seats to the smallest
country, *is impossible*.

3.2 The Quota Adjustment with Parabolic Functions

We now attempt an adjustment of the kind of $A(x) = a + bx + cx^2$ (which
we call parabolic); it is more flexible that the linear one because it has one more
parameter than the rectilinear adjustment, and it allows for the present situation
(and for a wide range of values of H) to assign six seats to the smallest country,
and 96 to the most populated one (using Webster or Jefferson for the rounding).
Also, as we show later, we can use this kind of adjustment analogously, facing
future enlargements of the EU.

For simplicity, in the following, as occurs at present with seven countries
having quotas under six, we suppose $m \leq 6$.

Let f_1 and f_2 be the following functions:

$$f_1(x) = 6 + \frac{90}{M-m}(x - m)$$
$$f_2(x) = 6 + \frac{90}{M-m}(x - m) - \frac{90}{(M-m)^2}(x - m)(x - M)$$

Their graphs (in Figure 1) are the straight line connecting the points $(m, 6)$ with $(M, 96)$ and the parabola connecting the same points and slope zero at $x = M$.

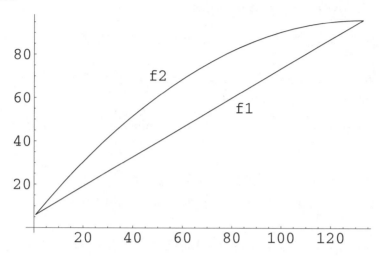

Fig. 1. Graph of the functions f_1 (straight line) and f_2 (parabola).

Now, rounding with the Webster method, we denote:

$$r_1 = \sum_{i=1}^{n} [f_1(x_i)]_W \text{ and } r_2 = \sum_{i=1}^{n} [f_2(x_i)]_W,$$

For example, for the present data, we have $r_1 = 635$ and $r_2 = 846$.

We define the *parabolic adjustment function* for the Webster method as the following:

$$A_\lambda(x) = \begin{cases} 6 + \lambda(x - m) & \text{if } H < r_1, \\ 6 + \frac{90}{M-m}(x - m) + \lambda(x - m)(x - M) & \text{if } r_1 \leq H \leq r_2, \\ 96 + \lambda(x - M)^2 & \text{if } H > r_2, \end{cases}$$

where, in each case, the λ parameter is used to adjust the allotment to H seats (that is, to verify the fourth criterion).

In the first case, the smallest country obtains six seats. In the second one, the smallest country obtains six seats and the largest 96 seats. And in the final case, the largest country obtains 96 seats.

In the following, we say that the value of H is *compatible with the bounds* of the projected Constitution when $6n \leq H \leq 96n$.

PROPOSITION 1 *If the value of H is compatible with the bounds, there always exists, at least one value for λ for which an allotment obtained, applying* *the Webster method to the parabolic adjustment function given above, will verify the four established criteria.*

Proof. Trivially, the value for λ must be

i) In the first case, $0 \le \lambda < \frac{90}{M-m}$, and the Webster method allots between $6n$ and r_1 seats.

ii) In the second case, $-\frac{90}{(M-m)^2} \le \lambda \le 0$, with the Webster method allotting between r_1 and r_2 seats.

iii) In the third case, $\lambda < 0$, permits the allotment of up to $96n$ seats.

∎

PROPOSITION 2 *We can obtain r_1 and r_2 for the rounding with the Jefferson method ($r_1 = 626$, $r_2 = 833$). Then, a similar proposition is true for the corresponding A_λ.*

REMARK 3 *We have used f_1 and f_2 connecting the points $(m, 6)$ and $(M, 96)$ because this permits using either the Webster or Jefferson method and assigns 6 seats to the smallest country and 96 to the biggest country. But if we establish just one method for rounding, for example Webster, we can change the 6 to any value from the interval (5.5, 6.5) and the 96 to any other belonging to (95.5, 96.5), etc. In (Ramirez, 2004) and (Ramirez, Palomares and Marquez, 2006) we use only the Webster method for rounding and we show the results obtained changing the interpolation point $(m, 6)$ to $(0, 5.5)$.*

The parabolic method applied to the present data in the European Parliament.

For the present data, the adjustment function is like the second one above, that is

$$A_\lambda(x) = 6 + \frac{90}{M - m}(x - m) + \lambda(x - m)(x - M)$$

whether we use Webster or Jefferson for the rounding.

When Webster is used for the rounding, the possible values of λ are those that belong to the interval

$$\lambda \in [-0.0023144, -0.0023140].$$

And when Jefferson is used the possible values of λ are those that belong to the interval

$$\lambda \in [-0.0026590, -0.0026583].$$

The allotments are shown in Table 1 (columns 5 and 7, both in bold). Also in this table we can see the inhabitants, the exact quota, the adjusted quota

Table 1. Current data for the EU and parabolic allotment, using W and J methods.

Country	Inhabitants	Exact Quota x	W. Adjust. $A_\lambda(x)$	Parabolic W-25	J. Adjust. $A_\lambda(x)$	Parabolic J-25	Present
Germany	82536700	132.92	96.00	96	96.00	96	99
France	59630100	96.03	79.04	79	80.25	80	78
United K.	59328900	95.54	78.78	79	80.00	80	78
Italy	57321000	92.31	76.98	77	78.26	78	78
Spain	41550600	66.91	61.21	61	62.72	62	54
Poland	38218500	61.55	57.50	57	58.99	58	54
Netherlands	16192600	26.08	29.59	30	30.53	30	27
Greece	11018400	17.74	22.20	22	22.87	22	24
Portugal	10407500	16.76	21.30	21	21.94	21	24
Belgium	10355800	16.68	21.22	21	21.86	21	24
Czechia Rep.	10203300	16.43	21.00	21	21.63	21	24
Hungary	10142400	16.33	20.91	21	21.54	21	24
Sweden	8940800	14.40	19.13	19	19.69	19	19
Austria	8067300	12.99	17.83	18	18.34	18	18
Denmark	5383500	8.67	13.77	14	14.11	14	14
Slovakia	5379200	8.66	13.37	14	14.10	14	14
Finland	5206300	8.38	13.50	14	13.83	13	14
Ireland	3963600	6.38	11.59	12	11.83	11	13
Lithuania	3462600	5.58	10.81	11	11.03	11	13
Latvia	2331500	3.75	9.05	9	9.18	9	9
Slovenia	1995000	3.21	8.52	9	8.63	8	7
Estonia	1356000	2.18	7.52	8	7.58	7	6
Cyprus	715100	1.15	6.50	7	6.52	6	6
Luxembourg	448300	0.72	6.08	6	6.08	6	6
Malta	397300	0.64	6.00	6	6.00	6	5
total	454552300	732	729.80	732	743.50	732	732

for $\lambda = -0.002314$ when the Webster method is used (A_W), and for $\lambda = -0.0026590$ when Jefferson is used (A_J), and the present allotment for the current 25 member States of the EU.

As above, the value of λ is not unique (except when there are ties).

Graphs a), of Figure 2, show the parabola of quota adjustment and the allotment with the parabolic method rounding with J. Graph b), shows the same parabola and the present allotment.

A similar view can be obtained when we compare the parabolic allotment, using Webster for rounding, with the present allotment.

Application of the parabolic method facing enlargements in the EU.

Let us suppose that the European Parliament is fixed in the maximum allowed, $H = 750$, and that we consider the following enlargements of the present EU, that is the EU-25:

a) EU-25 + Bulgaria + Romania (noted by EU-27).

b) EU-25 + Bulgaria + Romania + Croatia (noted by EU-28).

c) EU-25 + Bulgaria + Romania + Croatia + Turkey (noted by EU-29).

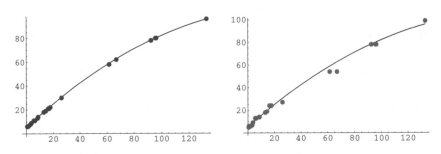

Fig. 2.   a) Parabolic adjustment function, and corresponding allotment, rounding with Jefferson for 25 member States. b) Parabolic adjustment function with Jefferson, and present allotment.

In these cases the corresponding values of r_1 and r_2 are:

	Rounding with W			Rounding with J		
	EU-27	EU-28	EU-29	EU-27	EU-28	EU-29
r_1	680	690	768	670	680	758
r_2	917	932	1025	903	917	1009

The corresponding allotments using parabolic adjustment appear in Table 2.

The quota adjustment function used in each case is the following:

EU	Round	Parabolic adjustment function
EU-27	W	$6 + \frac{90}{127.48 - .61365}(x - .61365) - .001629(x - .61365)(x - 127.48)$
EU-27	J	$6 + \frac{90}{127.48 - .61365}(x - .61365) - .0019(x - .61365)(x - 127.48)$
EU-28	W	$6 + \frac{90}{126.33 - .60809}(x - .60809) - .00143(x - .60809)(x - 126.33)$
EU-28	J	$6 + \frac{90}{126.33 - .60809}(x - .60809) - .0017(x - .60809)(x - 126.33)$
EU-29	W	$6 + .78538(x - .53543)$
EU-29	J	$6 + .802(x - .53543)$

Figure 3, graph a) shows the parabolic quota adjustment function and corresponding allotment for EU-27, rounding with Webster. Graph b) is analogous, but for EU-28, and rounding with Jefferson.

For EU-29, the two quota adjustment functions are straight lines; so, if the enlargements had affected more countries, the parabolic adjustment function would also be a line.

Table 2. Parabolic allotments for enlargements of the EU, using W and J methods.

State	Inhabitants	W-27	J-27	W-28	J-28	W-29	J-29
Germany	82536700	96	96	96	96	93	94
Turkey	66500000	-	-	-	-	76	77
France	59630100	76	77	75	76	69	70
United K.	59328900	76	76	75	76	69	69
Italy	57321000	74	74	73	74	66	67
Spain	41550600	58	58	57	57	50	50
Poland	38218500	54	55	53	54	46	46
Romania	22000000	35	36	35	35	30	29
Netherlands	16192600	27	28	27	27	23	23
Greece	11018400	21	21	20	20	17	17
Portugal	10407500	20	20	19	19	17	16
Belgium	10355800	20	20	19	19	16	16
Czechia Rep.	10203300	20	19	19	19	16	16
Hungary	10142400	19	19	19	19	16	16
Sweden	8940800	18	18	17	17	15	15
Bulgaria	8428000	17	17	17	17	14	14
Austria	8067300	17	16	16	16	14	14
Denmark	5383500	13	13	13	12	11	11
Slovakia	5379200	13	13	13	12	11	11
Finland	5206300	13	12	13	12	11	11
Croatia	4436000	-	-	11	11	10	10
Ireland	3963600	11	11	11	11	10	9
Lithuania	3462600	10	10	10	10	9	9
Latvia	2331500	9	8	9	8	8	8
Slovenia	1995000	8	8	8	8	8	7
Estonia	1356000	7	7	7	7	7	7
Cyprus	715100	6	6	6	6	6	6
Luxembourg	448300	6	6	6	6	6	6
Malta	397300	6	6	6	6	6	6
total	454552300	750		750		750	

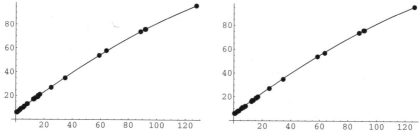

Fig. 3. a) Parabolic quota adjustment function, and allotment, rounding with W for
EU-27 and $H = 750$. b) The same for EU-28, and rounding with Jefferson.

3.3   The Quota Adjustment with Power-Type Functions

Now, for the r_1 value obtained previously using Webster, we consider the
following function for the adjustment of the quotas, defined in $[m, M]$:

$$A_\alpha(x) = \begin{cases} 6 + \frac{90\alpha}{M-m}(x - m) & \text{if } H \leq r_1, \\ 6 + 90\left(\frac{x-m}{M-m}\right)^\alpha & \text{if } H > r_1 \end{cases}$$

Table 3. Power type allotments for the present data in EU-25 and for several enlargements of the EU.

State	Inhabitants	Present	W-25	J-25	W-27	J-27	W-28	J-28
Germany	82536700	99	96	96	96	96	96	96
France	59630100	78	75	75	74	74	73	73
United K.	59328900	78	75	75	73	73	73	73
Italy	57321000	78	73	73	71	71	71	71
Spain	41550600	54	58	58	55	56	55	55
Poland	38218500	54	54	54	52	52	51	51
Romania	22000000	-	-	-	35	35	34	35
Netherlands	16192600	27	30	30	28	28	27	27
Greece	11018400	24	23	24	21	21	21	21
Portugal	10407500	24	23	23	20	21	20	20
Belgium	10355800	24	23	23	20	20	20	20
Czechia Rep.	10203300	24	22	22	20	20	20	20
Hungary	10142400	24	22	22	20	20	19	20
Sweden	8940800	19	21	21	19	19	18	18
Bulgaria	8428000	-	-	-	18	18	17	18
Austria	8067300	18	19	19	18	17	17	17
Denmark	5383500	14	15	15	14	14	13	13
Slovakia	5379200	14	15	15	14	14	13	13
Finland	5206300	14	15	15	14	14	13	13
Croatia	4436000	-	-	-	-	-	12	12
Ireland	3963600	13	13	13	12	12	11	11
Lithuania	3462600	13	12	12	11	11	11	11
Latvia	2331500	9	10	10	9	9	9	9
Slovenia	1995000	7	10	10	9	9	9	8
Estonia	1356000	6	9	8	8	8	8	7
Cyprus	715100	6	7	7	7	6	7	6
Luxembourg	448300	6	6	6	6	6	6	6
Malta	397300	5	6	6	6	6	6	6
total	454552300	732	732	732	750	750	750	750

(When $H > r_1$ and $\alpha = 0$ we establish $A_\alpha(m) = 96$)

In both cases, when $\alpha \in [0, 1]$ the function A_α is increasing, concave, with values belonging to $[6, 96]$, then we can use the parameter α to adjust the sum to H.

PROPOSITION 4 *If the value of H is compatible with the Constitutional bounds, there exists always, at least one value of α for which an allotment obtained applying the Webster method to the power-type adjustment function above will verify the four criteria for degressively proportional allotment.*

Proof. If $H \leq r_1$, $A_\alpha(x) = 6 + \frac{90\alpha}{M-m}(x - m)$, and then varying α from 0 to 1, we can obtain all the allotments that sum from 6n to r_1.

If $H > r_1$, the adjustment quota function is $A_\alpha(x) = 6 + 90 \left(\frac{x-m}{M-m}\right)^\alpha$ so when α decreases, A_α increases, and $\lim_{\alpha \to 0} A_\alpha(x) = 96, \forall x \in [m, M]$; then using Webster, all countries can obtain 96 seats. ∎

PROPOSITION 5 *If we use the Jefferson method to calculate r_1 and to round the adjusted quotas, a very similar affirmation to Proposition 3 can be established and proofed.*

The power method for the present situation in the European Parliament and some enlargements.

For the present 25 members of the EU with $H = 732$ and the enlargements EU-27, EU-28 with $H = 750$, the corresponding adjustment functions are:

EU	H	Rounding	Adjustment function
EU-25	732	W	$6 + 90 \left(\frac{x - .63980}{132.92 - .63980} \right)^{.803}$
EU-25	732	J	$6 + 90 \left(\frac{x - .63980}{132.92 - .63980} \right)^{.786}$
EU-27	750	W	$6 + 90 \left(\frac{x - .61365}{127.48 - .61365} \right)^{.8675}$
EU-27	750	J	$6 + 90 \left(\frac{x - .61365}{127.48 - .61365} \right)^{.85}$
EU-28	750	W	$6 + 90 \left(\frac{x - .60809}{126.33 - .60809} \right)^{0.892}$
EU-28	750	J	$6 + 90 \left(\frac{x - .60809}{126.33 - .60809} \right)^{0.8655}$

The corresponding allotments for the present EU-25 as well as the EU-27 and EU-28 using W and J, and using this power-type method, are shown in Table 3. For EU-29, the parabolic quota adjustment function and the power one are the same straight line. Therefore, the power-type allotment for EU-29 is the same as the parabolic method.

In the graph of Figure 4, we compare the power-type methods, rounding with J, with the present allotment.

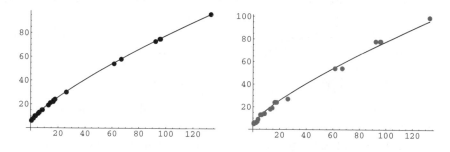

Fig. 4. a) Power-type adjustment function, and corresponding allotment, rounding with Jefferson for EU-25. b) Power-type adjustment function with Jefferson, and present allotment.

In the graph of Figure 5, we compare parabolic adjustment quota function and power-type adjustment quota function, and corresponding allotments, rounding with Webster, for EU-27.

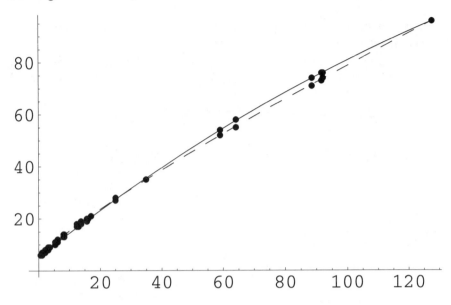

Fig. 5.   Parabolic adjustment quota function (continuous) and power-type adjustment quota function (broken line), and corresponding allotments, rounding with Webster, for EU-27.

REMARK 6 *The power-type adjustment function generally grows faster than the parabolic one when the quota x is near m, and the contrary occurs when the quota x is near M. Therefore, the power-type adjustment method using Webster is more favourable for the small countries, and the parabolic method using Jefferson favours the larger countries.*

We can see this fact comparing Table 3 with tables 1 and 2; and with the graphs of figures 4 and 5 we show these differences for EU-27 and EU-28.

3.4   Other Quota Adjustment Functions

Many other adjustment functions can be proposed, usually more complex than the parabolic method or the power-type method shown above.

For example splines functions. Linear splines suffice, even with just three nodes: $\{m, k, M\}$; k being an intermediate node between m and M. If $r_1 < H$, then we use the spline connecting the points $(m, 6)$, (k, b) and $(M, 96)$. There is always a linear spline function with three nodes for the quota adjustment that verifies the four properties, and we choose the b parameter such that one of the roundings sum up to H. Besides, there are many valid values for the k intermediate node. Precisely, this versatility becomes a disadvantage for using the splines. What intermediate node do we use? (This is even more complex if we increase the number of nodes.)

The spline is not differentiable in its intermediate node, which is another disadvantage with respect to the parabolic method and power-type method. Splines involve a more abrupt change than the one resulting from these two methods. Furthermore, for the EU of 29 countries, as well as any future enlargements, the spline method, the parabolic method and the power-type method produce the same quota adjustment function, a straight line, and so they produce the same allotment.

4.   What Method to Use?

We have developed parabolic and power-type adjustment functions rounding with Webster and Jefferson, four total possibilities. All of them are logical options for solving the problem at hand. When differences are seen, they are small, in some cases. For the EU-25, the bigger differences can be seen in Figure 6, where we compare the power-type method, rounding with Webster, versus the parabolic method, rounding with Jefferson.

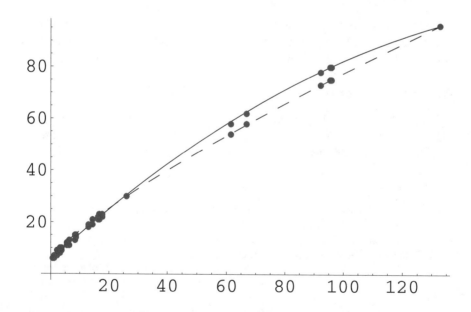

Fig. 6.   Parabolic quota adjustment function rounding with Jefferson (continuous) and power-type quota adjustment function rounding with Webster (broken line), and corresponding allotments for EU-25.

Also, we observe:

- The adjustment of the parameter is of the same difficulty whether we use the parabolic or the power-type method. In both cases it is easy.

  For the rounding method chosen, we have to calculate r_1 and r_2, and then depending on these values, and H, we will know the expression of the A_λ or A_α function. Next a method similar to bisection is used to easily obtain a value of the parameter that allots the H seats. Only in the case where more than one solution exists (that is, a tie), minor additional work is required.

  In short, both the parabolic method and the power-type method, are very easy to apply.

- While Webster's is an impartial method and Jefferson favours the bigger countries, as the adjusted quotas are between 6 and 96 this causes that the differences obtained between using Webster or Jefferson are not important. As we can see in tables 1, 2 and 3, the difference between using Webster or Jefferson method is always at most one seat, for every country.

- Power-type methods are most favourable than the parabolic one for small countries.

Therefore, if for example, as the projected Constitution gives a minimum of 6 seats to the smaller countries, the parabolic method is the method that more compensates this Constitutional restriction better than the power-type one.

Acknowledgments

The authors thank the '*Junta de Andalucia*' for its support through group FQM-191. We also thank the referee for relevant comments and suggestions that greatly improved this paper.

References

Balinski, Michel L., and Hobart P. Young. (1982). *Fair Representation: Meeting the Ideal of One Man One Vote*. New Haven, CT.

Ramirez, Victoriano. (2004). *Some Guidelines for an Electoral European System*, Workshop on Institutions and Voting Rules in the EC. Sevilla (Spain).

Ramirez V., Palomares A., and Marquez M.L.. (2006). Un metodo para distribuir los escaños del parlamento Europeo entre los Estados miembros de la UE, *Revista de Ciencia Politica*, 15.

Tratado por el que se establece una Constitución para Europa, Ministerio de Asuntos Exteriores y de Cooperación, Ministerio del Interior y Ministerio de la Presidencia, Dep. Legal M-53614/2004.

Hidden Mathematical Structures of Voting[*]

Donald G. Saari
University of California, Irvine

Abstract   The complexities of voting theory are captured by Arrow's Impossibility Theorem and McKelvey's chaos result in spatial voting. A careful analysis of Arrow's theorem, however, proves that not all of the supplied information is used by the decision rule. As such, not only does this seminal result admit a benign interpretation, but there are several ways to sidestep Arrow's negative conclusion. McKelvey's result is described in terms of more general voting rules. Then a new solution concept, called the 'finesse point', is introduced. This centrally located point generalizes the core and minimizes what it takes to respond to any proposal by another person.

Keywords: Arrow's theorem, spatial voting, majority vote, chaos theorem, core, finesse point.

1.   Introduction

For me, an important part of the excellence of this Erice conference came from the diverse intellectual interactions that allowed us to learn from others. In this spirit, I changed my conference presentation to respond to comments made in earlier talks. We heard, for instance, how the negative conclusion of Arrow's Theorem (Arrow, 1952), often described as showing that "no election rule is fair," proves it is impossible to solve the central problems of social choice and voting theory. Other conference comments were directed toward the distinct weaknesses of pairwise majority vote elections as manifested by the negative consequences of McKelvey's "chaos theorem" (McKelvey, 1979).

After reminding the reader what Arrow's and McKelvey's seminal theorems state, I explain why neither result is as discouraging as widely assumed. Instead, both results, which involve pairwise comparisons, can be handled without undue difficulty. As part of my explanation, I introduce the "finesse point" concept, which is a new way to extend the notion of a "core."

---

[*] My thanks to a referee who clearly did a careful job as he caught some subtle errors!

2.   Arrow's Theorem

After identifying the domain and range of a social welfare function, Arrow's seminal theorem describes conditions that most people probably would accept as being satisfied by all reasonable decision rules. In particular, for voter preferences, Arrow stipulates

> **Preferences:** Each voter has a complete, transitive ranking over all alternatives; there are no other restrictions on the voter's ranking.

Arrow also requires the outcomes to be transitive.

> **Societal Ranking:** The societal ranking forms a complete transitive ranking.

Transitivity is a tradition assumption in this area. This rationality condition is where if, say, a voter prefers Aline to Federica, and Federica to Susan, then he must also prefer Aline to Susan. A rationale for this assumption is to help ensure that societal decisions can be made. If, for instance, all the above voters had these preferences where they really did prefer Susan to Aline, then who do they want?

> Not Aline, because they prefer Susan. Not Susan because they prefer Federica. Not Federica because they prefer Aline.

So without requiring voters to have transitive beliefs, we must anticipate cyclic societal outcomes. To achieve transitive societal rankings, then, it is natural to require the voters to have transitive preferences. In other words, these two "transitivity" conditions are inextricably intertwined.

We now come to the conditions Arrow imposed on the decision rule. The first is obvious: if *everyone* prefers one candidate to another, then the unanimously accepted ranking should be the societal ranking. The second condition extends this unanimity notion to more general settings. For instance, suppose a committee ranking candidates for tenure track positions prefers Anne to Barb. Imagine Barb's reaction if told that the committee would have ranked her over Anne if they had a better opinion of *Connie!* Why should their opinions of Connie affect the {Anne, Barb} ranking? Shouldn't their opinions about Connie be irrelevant for the Anne, Barb ranking?

> **Decision rules.** *Pareto, or unanimity.* If for any pair of candidates all voters rank them in the same manner, then that common ranking is the societal ranking.
>
> *Binary Independence, or Independence of Irrelevant Alternatives (IIA).* The societal ranking of any two candidates strictly depends on the voters rankings of these two candidates; all other information is irrelevant. In particular, for any two profiles whereby each voter has the same relative ranking of a pair, the pair's societal ranking is the same.

These conditions sound quite reasonable and even innocuous. It is their reasonableness that constitutes the surprise and mystery of Arrow's Theorem.

THEOREM 1 *(Arrow, 1952) With three or more alternatives, any decision rule that satisfies the above conditions on preferences, the societal ranking, and the properties for a decision rule must be equivalent to a dictatorship. Namely, there is a specific voter whereby the societal outcome always agrees with that voter's preferences independent of the choices for any other voter.*

2.1   An Explanation

A dictator? How can that be!! The power of Arrow's result derives from selecting minimal conditions that most people probably assume hold for all decision rules. But Arrow proved that unless subjected to a dictatorship, some of these conditions must be violated. As these rules appear to be desirable, it is easy to appreciate why the mystery of Arrow's result helped to generate the rebirth of the academic area and why his theorem often is described as meaning "no voting rule is fair."

A half century later and after examining the hidden mathematical structures of Arrow's Theorem (Saari, 1998,2001), we now understand that Arrow's result does not mean what we had thought it meant. Instead, mathematics proves that Arrow's result admits a benign interpretation. Even more; we now can replace the difficulties of Arrow's negative result with positive conclusions.

Before indicating the source of Arrow's conclusion, let me mention that during this conference, I discovered, to my delight, that the Erice restaurants served Sicilian red wine, white wine, and beer. With enjoyable experimentation, I learned that I prefer their red wine to their white. The question is: over these three choices, are my preferences transitive?

This question is impossible to answer because there is not enough information. To determine whether my preferences are transitive, you must also know my {white wine, beer} and my {red wine, beer} rankings. This aside explains Arrow's result; when determining the societal ranking for Sicilian red and white wines, IIA, or the binary independence condition, forbids using any information about beer. More precisely, IIA requires the decision rule to *ignore any and all information about the transitivity of the election rankings of the individuals.* This observation is crucial because, as emphasized above, without using the transitivity of individual preferences, there is no reason to expect transitive societal outcomes. But the decision rule must ignore this crucial assumption, so the surprise and mystery of Arrow's theorem completely disappears. A more correct assertion is that *Arrow's theorem shows that serious negative consequences can arise by ignoring valuable information, which everyone expected was being used, to determine the societal ranking.*

Before explaining Arrow's dictator, permit me another digression to brag about my youngest granddaughter. At ten months, Tatjana could move along alphabet blocks on the floor by first crawling by the "A" block, then the "B" block, next the "C" block, and so forth in the correct order down the alphabet! Honest! Well I helped a bit by arranging the blocks in the appropriate order along her crawling path. OK, so her accuracy in selecting the blocks may not measure her acumen; instead it may register the careful ordering of the data. Similarly, because a rule obeying IIA is forbidden from using information about the transitivity of individual rankings, the rule needs significant help if it is to register a transitive societal outcome. To achieve this conclusion, mimic the alphabet block scheme by using highly selective data. But by doing so, remember that it is the structure of the data, not the properties of the decision rule, that *imposes* order on the societal outcome: this explains the role of "profile restrictions" in choice theory. An extreme condition is to use the transitive rankings of a single agent: this is Arrow's dictator. In other words, *rather than describing a decision rule, Arrow's "dictator" is an extreme version of a profile restriction.*

Mathematical Structure. For readers interested in what hidden mathematical structures allowed me to discover the source of Arrow's result, notice how the Pareto and binary independence requirements directly emphasize *pairs.* The main role played by the Pareto condition (Saari, 1995,1998) is to ensure that each pair has at least two societal rankings. Binary independence requires that the way the societal ranking of a pair is determined is independent of what happens with others.

These conditions decompose a ranking into its pairwise rankings; e.g., $A \succ B \succ C$ becomes $A \succ B$, $B \succ C$, and $A \succ C$. A change in a strict ranking of any pair is represented by an operation in the group Z_2; it either keeps the ranking, or it reverses it. To find what can happen over all three pairs, it follows from IIA that we must examine the $Z_2 \times Z_2 \times Z_2$ orbit of any ranking—this orbit has eight rankings, not just the six transitive rankings—the two new rankings are cyclic. By examining this structure of "completing the space" of individual preferences, the argument becomes immediate.

2.2   Finding Resolutions

The source of Arrow's Theorem, then, is that his conditions prevent the decision rule from using information that is explicitly specified. This means that rather than Arrow's conclusion that *the decision rule must be equivalent to a dictatorship,* a more accurate conclusion is that

> *Arrow's Theorem means that his conditions prevent decision rules from using the crucial information about individual rationality that we expect is being used.*

Arrow's theorem is not the only negative result with this explanation; as explained in my November, 2004, Condorcet Lectures (Saari, 2004), all of the impossibility or "troublesome" results that I have investigated (from social choice, probability, etc.) occur because the decision rule does *not* use crucial information that, by being specified, we expect is being used. (For instance, with respect to Sen's theorem, see Saari and Petron (2006) and Li and Saari (2005).) An important payoff from this description is that we now know how to avoid the negativity associated with all of these conclusions: modify the conditions so that the rules *can* use the information that we find to be vital.

In Arrow's result, the approach is simple. Modify IIA, or binary independence, in a manner so that the decision rule *can* use the crucial information about the transitivity of voter preferences. While all of this is described in my book Saari (2001), a simple approach is to not only rank each pair, but also specify how many alternatives separate the pair. For instance, the $A \succ B \succ C$ ranking is treated as $(A \succ B, 0)$, $(B \succ C, 0)$, and $(A \succ C, 1)$ where the "0" or '1' indicates how many candidates separate the specified ranking. With this modified "intensity IIA condition" (IIIA), Arrow's dictator is replaced by several rules including the Borda Count.

3.   McKelvey's Chaos Theorem

Pairwise majority vote elections are widely used in departmental meetings, legislatures, and on and on. But what do they mean? Arrow's theorem suggests that we must expect negative conclusions, and they most surely occur. In Saari and Sieberg (2001), for instance, we show that rather than viewing pairwise outcomes as reflecting the actual profile, they actually reflect *a statistical average over all possible supporting profiles.* Notice how this interpretation almost ensures all sorts of unintended and inappropriate outcomes! To illustrate this phenomenon, Sieberg created a clever example for Saari and Sieberg (2001) to show how the same election outcome can support radically different intentions and interpretations. In Saari (2006) I build on her example to develop a different explanation that shows how the pairwise vote, or IIA, severs all intended connections among pairs.

In spite of its many faults and weaknesses, the pairwise vote continues to be used. As such, it remains reasonable to discover ways to temper the negativity with positive conclusions, and that I will do. Start with a single issue where each voter has an "ideal point"—a preferred position: propositions closer to her ideal point are more preferred. If the positions of three voters are indicated with bullets in Fig. 1a, where should a candidate position herself? One choice might be to take the average position; another choice might be to assume the midpoint between the extremes, as indicated by the arrow representing candidate 1.

Neither choice is optimal. The optimal choice, as selected by candidate 2 in Fig. 1a, is the 'median voter's' position. By being at this point, candidate 2 earns the votes of the median voter and the one to the left; candidate 1 loses by obtaining only the vote of the voter to the right. Experimentation proves that no matter where candidate 1 is positioned, candidate 2 will win.

a. Three voters   **b.** Four voters

Fig. 1.   What policies to adopt?

Candidate 2's position illustrates the game theoretic notion of a "core." More precisely, *a core point is one that can never be beaten under the rules of the game:* the 'core' is the set of all core points. But while a core point can never be beaten, it need not win. An illustration is in Fig. 1b where the median—the core—is the interval between the second and third voters' ideal points. With both candidate's positions in the core, neither can be beaten, but neither wins. It is easy to see that with one issue (where the modeling is on a line), a core always exists; this is, essentially, Black's single peaked condition. Now consider what happens with two issues, which is modeled in a two-dimensional setting.

3.1   Two Issues, Two Dimensions

Consider a simple setting where, say, a committee of three must recommend the number of hours and salary paid to teaching assistants. A voter's ideal point, then, consists of two coordinates (hours, salary). If the three voters' ideal points define a triangle, what should be the compromise 'decision point'? Most surely the group would never accept a point such as the dagger to the far right of the Fig. 2a triangle; or would they? Presumably agreement is somewhere in the center of the triangle; perhaps where the three dashed circles in Fig. 2a meet.

This central point is *not* a core point. The proof is immediate: each circle's center is a voter's ideal point, so the voter prefers anything inside of the circle to the proposal. These three circles intersect in a trefoil: each leaf of the trefoil indicates choices preferred by a winning coalition to the purposed choice. For example, anything in the large lower leaf can be achieved by the coalition $\{1, 2\}$. Similarly, anything in the smaller leaf on the left can be achieved by coalition $\{1, 3\}$.

Indeed, the geometry of circles ensures that unless the ideal points for the three voters are on a line, then *for any proposal, there always is a counter-proposal that can win by being supported by a two voter coalition.* Thus the majority vote core for three voters and two issues does not exist.

A reaction might be, so what? What difference does it make whether the core exists? The consequences for democracy are startling and surprisingly

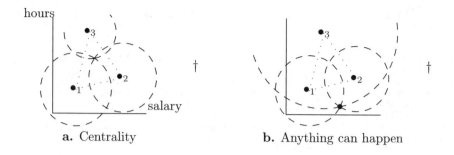

a. Centrality   **b.** Anything can happen

Fig. 2. More issues

negative: this is the McKelvey chaos theorem, which asserts that a society can start from any specified position and, even with sincere majority votes, end up at any other specified position. Illustrating with Fig. 2a, McKelvey's result proves it is possible to start at the center of this circle, and, with an appropriately designed agenda, these voters will sincerely adopt the dagger that is far from any of their ideal points! After stating this beautiful theorem, I will indicate how surprisingly easy it is to achieve this outcome.

THEOREM 2 *(McKelvey, 1979) For any number of voters, suppose that the majority vote core for their ideal points is empty. It is possible to specify starting and ending positions $\mathbf{p}_s$, $\mathbf{p}_f$ and an agenda—a list of proposals $\{\mathbf{p}_j\}_{j=1}^N$ so that $\mathbf{p}_1 = \mathbf{p}_s$ and $\mathbf{p}_N = \mathbf{p}_f$ and $\mathbf{p}_{j+1}$ will beat $\mathbf{p}_j$, $j = 1, \ldots N - 1$, in a majority vote.*

To illustrate McKelvey 'chaos theorem' with the Fig. 2a challenge, I will indicate how to start at the central point and design an agenda so that the dagger will be accepted by these voters. At the first stage let voters 1 and 2 combine to select a point near the bottom of the Fig. 2a leaf in the trefoil; e.g., the bullet below the triangle in Fig. 2b. Redraw the circles with this new proposal—points in the larger circle with voter three's ideal point at the center and that for voter two now come much closer to the dagger. It should be clear that creating an agenda to reach the dagger is fairly easy—even though all voters prefer the center point to the dagger!

4. Challenges

McKelvey's theorem raises at least two mathematical challenges:

1 When does the core exist? More precisely, as the core always exists for one issue, what is the maximum number of issues for which the core will exist?

2 When the core fails to exist, can we find a substitute concept to replace core points?

To provide a more complete exposition, I answer both questions for "q-rules." This *quota* rule is where, to win in a pairwise comparison, you need at least q of the n possible votes where $\frac{n}{2} < q \leq n$: denote the q-rule core by $\mathbb{C}(q, n)$. The lower bound, where q is the first integer greater than $\frac{n}{2}$, is the majority vote, while the upper bound, where $q = n$, is the unanimity rule. As $\mathbb{C}(n, n)$ is the convex hull of the n ideal points, it always exists.

By increasing the quota (i.e., selecting a larger q value), the stability (perhaps of the status quo) also increases. By this I mean that if $q_1 < q_2$ and $\mathbb{C}(q_1, n)$ exists, then $\mathbb{C}(q_2, n)$ exists and $\mathbf{C}(q_1, n) \subset \mathbb{C}(q_2, n)$. (For a proof, if $\mathbf{p} \in \mathbb{C}(q_1, n)$, then no proposal can get q_1 votes against $\mathbf{p}$. In turn, no proposal can obtain $q_2 > q_1$ votes against $\mathbf{p}$, so $\mathbf{p} \in \mathbb{C}(q_2, n)$.) With this added stability, we must wonder whether McKelvey's result extends to other q-rules. It does.

My former student Monica Tararu (Tataru, 1996,1999) proved in her Ph. D. thesis that if $\mathbb{C}(q, n) = \emptyset$, then McKelvey's conclusion, where it is possible to start anywhere and end up anywhere else, holds for this q-rule. She did more: she established upper and lower bounds on the minimum number of items needed in an agenda to accomplish this behavior. Her bounds use the distance between the specified starting and ending position divided by a term determined by the ideal points. But the disquieting sense of electoral instability promoted by the "chaos theorem" (McKelvey's and Tataru's results) suggests augmenting the above mathematical challenges with the following:

3. Find a "natural stability" concept that counters the chaos theorem's sense of instability.

4.1   Plott Diagrams

C. Plott (1967) proved for any number of issues that ideal points can be positioned so that a majority vote core exists. Because $q_1 < q_2$ requires $\mathbb{C}(q_1, n) \subset \mathbb{C}(q_2, n)$, Plott's diagrams also establish the existence of $\mathbb{C}(q, n)$, $q > \frac{n}{2}$, for any dimensional space. An example is Fig. 3a, which shows that the key to the Plott diagrams is symmetry. To see why voter 5's ideal point is the Fig. 3a core, select any other proposal, such as the diamond. As illustrated by the dashed lines, draw a line (the horizontal one) perpendicular to the line from voter 5's ideal point to the diamond. Voter 5's point wins as it is supported by the majority $\{1, 4, 5\}$ who are on or below the horizontal line.

To understand Plott's approach, start with a setting where a core point exists; e.g., with points on the line. Rotate pairs of points about the core point as indicated by the Fig. 3a dotted curves with points 1 and 3. The symmetry of the construction allows the core point to persist, but precariously. More

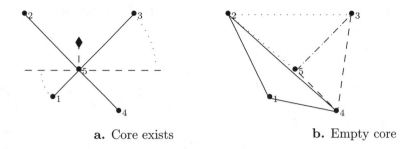

a. Core exists   **b.** Empty core

Fig. 3.   Plott diagram and sensitivity of the core

precisely, without this exacting symmetry, $\mathbb{C}(3,5)$ can vanish. In Fig. 3b, for instance, moving voter 2's ideal point ever so slightly to the left leads to an empty core. To see this, notice that the *Pareto set for a coalition* (i.e., points that if moved will result in a less preferred position for some coalition member) is the convex hull of coalition members' ideal points. So if $\mathbb{C}(3,5) \neq \emptyset$ for Fig. 3b, $\mathbb{C}(3,5)$ would be in the convex hulls for coalitions $\{2,3,5\}$ and $\{3,4,5\}$; i.e., $\mathbb{C}(3,5)$ would be on the line connecting points 3 and 5. But by disturbing the symmetry by slightly moving voter 2's ideal point, the $\{1,2,4\}$ Pareto set misses this line resulting in $\mathbb{C}(3,5) = \emptyset$.

The lesson learned is that while $\mathbb{C}(q,n)$ might exist, it could be useless if even the slightest change in preferences could cause it to vanish. Consequently, rather than asking whether $\mathbb{C}(q,n)$ exists, the above challenge 1 should be replaced with the more realistic and relevant challenge:

1'. When does the core exist *generically*? That is, for what dimensional spaces—for how many independent issues—will the the core persist even with small changes in preferences?

5.   Generic Existence of a Core

The quest to determine the dimensions of issue space for which $\mathbb{C}(q,n)$ can exist generically has a long history where, among several others, major contributions were made by Schofield (1963), McKelvey (1986), McKelvey and Schofield (1987), and Banks (1995); I finally resolved the issue in Saari (1997). (An exposition is in Saari (2004).)

It is reasonable, of course, to question why in the above description the voter preferences are given by circles. They need not be; in my resolution of the core problem, I replaced Euclidean distances with smooth utility functions. Thus small changes in preferences go beyond changing the ideal point to also permit small changes in the utility function. For instance, "small changes" permit the above circles to be replaced with ellipses, or other geometric objects.

The general answer for challenge 1' is given by the following theorem. Brace yourself; the statement is ugly.

THEOREM 3 *(Saari, 1997) Let k be the dimension of issue space. The core $\mathbb{C}(q,n)$ exists generically for $k = 1$. The core $\mathbb{C}(q,n)$ exists generically for $k \leq 2$ when $q = 3$ and $n = 4$. If $n \geq 5$ and $4q < 3n + 1$, then $\mathbb{C}(q,n)$ exists generically if and only if*

$$k \leq 2q - n. \tag{1}$$

For super-majorities in which $4q \geq 3n + 1$, let α be the largest odd integer such that $\frac{q}{n} > \frac{\alpha}{\alpha+1}$. $\mathbb{C}(q,n)$ exists generically if and only if

$$k \leq 2q - n - 1 + \frac{\alpha - 1}{2}. \tag{2}$$

The precise statement (Saari, 1997,2004) is even more complicated as it divides core points into categories where the k values differ. Also, the complicating α values reflect changes permitted by utility functions, so this term plays no role with the preferences described earlier.

Let me offer a common sense interpretation that holds for the earlier Euclidean preference discussion where changes are in the position of ideal points. In this setting, $\mathbb{C}(q,n)$ exists generically iff $k \leq 2q - n$. There is a simple way to appreciate this bound: the maximum number of issues equals the number of voters that must change views to convert a previously losing position into a winner. To illustrate, if $n = 100$ and $q = 80$, then there are 20 voters on the losing side. For this group to become winners, they must persuade $80 - 20 = 60$ voters to change their vote. Thus, for this example, $60 = 2q - n = 2(80) - 100 \geq k$ is the maximum number of issues.

An often expressed concern is to explain why actual political settings appear to be more stable than suggested by theory. An answer is suggested by how the rule for the bound on k is described. For instance, consider a close majority vote election in a city of the size of Minneapolis, where the winner might have only 50.2% of the vote given by the $502,000$ to $498,000$ tally. With an abuse of Thm. 3, notice how the pneumonic showing that the maximum k value is the number of voters that must change views to change the outcome also suggests that the election will remain stable for up to 2001 issues: this constitutes considerable stability. Moreover, the proof of Thm. 3 used the mathematical tool of "singularity theory," which carries an implicit assumption that each voter has views independent of others. This is not true in general: more realistic assumptions lead to a much stronger sense of stability—at the expense of a more complex mathematical analysis.

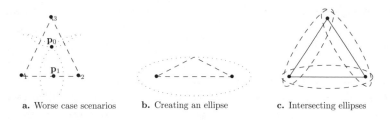

a. Worse case scenarios   **b.** Creating an ellipse   **c.** Intersecting ellipses

Fig. 4.   Finding the d-finesse point

6.   The Finesse Point

The core ensures stability while its absence suggests, via the chaos theorems, instability. It is natural to wonder whether a solution concept could be found between these extremes, where exercising thoughtful control ensures some level of stability. To understand the objective, if a core point cannot be beaten, then the sought after solution concept should be one that minimizes what needs to be done (in changing positions) to keep from being beaten. This is the *finesse point*.

To understand what it takes to avoid being beaten, we first need to understand how to defeat a position. So consider Fig. 4a and proposal p_0 at the bullet near the center. All points that can beat p_0 are in the trefoil, or "winning set," defined by the three circles passing through p_0. While any point in the winning set can defeat p_0, the counter-response is immediate. Suppose, for instance, that the challenge is given by point p_1 in the lower leaf. A response is to sufficiently change p_0 so that p_1 is outside the new winning set. This is easy to do. As p_1 is slightly closer to the circle with the first voter's ideal point as a center (call this distance α), an effective counter-response is to modify p_0 by moving α units directly toward voter 1's ideal point. The move reduces the circle's radius of the first circle by α, so the new winning set excludes the challenge put forth by p_1.

As this description indicates, the worse case scenario is if the p_1 challenge is located at the middle of widest point of the widest leaf; in Fig. 4a, this is where the bottom leaf intersects the edge connecting ideal points 1 and 2. Using the geometry of circles (and where the circles intersect on the $\{1, 2\}$ edge), the width of this leaf is $2d_{1,2} = r_1 + r_2 - z_{1,2}$, where r_j is the radius of the circle with center at the jth voter's ideal point that passes through p_0 and $z_{j,k}$ is the edge length between the jth and k^{th} ideal points. Here the maximum change in p_0 to respond to a counterproposal is $d_{1,2}$. Rewriting this equation in terms of $r_1 + r_2$, which is the sum of the distances from p_0 to the two ideal points, we have

$$r_1 + r_2 = z_{1,2} + 2d_{1,2} \tag{3}$$

More generally, if the sum of distances from *any proposed point* to the two ideal points satisfies Eq. 3, the width of the defined leaf is $2d_{1,2}$ and the maximum required change is $d_{1,2}$.

The above comments indicate how to find what I call the "*d*-finesse point." To see the basic idea, recall that an ellipse can be drawn by looping a string over two nails as indicated in Fig. 2b. With the looped string pulled taut with a pencil, move the pencil all the way around to draw an ellipse. With a string length $2z_{1,2} + 2d$, call the figure a "*d*-ellipse." The part of the string connecting the nails is $z_{1,2}$, so the portion of the string tracing out the figure has length $z_{1,2} + 2d$. Comparing this description with Eq. 3, we discover that *any point on a d-ellipse defines a winning set leaf with maximum width 2d*.

If a player has no knowledge about the nature of possible counterproposals to an initial $\mathbf{p}_0$, he should minimize the amount of change required to respond to *any* winning coalition. This requires finding a proposal $\mathbf{p}_0$ where the width of each winning set agrees. (If not so, then an opponent might select a point in a wider leaf.) Finding such a point is easy for three voters with the majority vote. About each pair, construct a *d*-ellipse. As indicated in Fig. 4c, there are three *d*-ellipses where any two intersect. At this intersection point, the maximum width of the leaf for each of the two coalitions is $2d$. If the third *d*-ellipse does not include this point, the maximum width (maximum change for a successful counterproposal) for the third leaf is larger. Thus, to minimize all possible worse case scenarios, we need to find the point where all three *d*-ellipses meet.

DEFINITION 4 *For three voters and the majority vote, the d-finesse point is where the three d-ellipses intersect for the minimal value of d. For a q-rule with more voters and issues and a winning coalition C, let $E_d(C)$ be the set consisting of the Pareto set of C and the d-ellipsoid for each pair of points in C. The d-finesse point is a point that is in $E_d(C)$ for all winning coalitions C, and d is the minimal value for which this is true.*

By construction, the *d*-finesse point is the location that requires a minimal response for any counterproposal; the value of d could be, but need not be, small. Let $\mathbb{S}_d(q)$ be the "adjustment region" where all adjustments to a position with a *q*-rule to regain a winning position are possible: $\mathbb{S}_d(q)$ is a sphere of radius d. Many results, including a description about how to find the *d*-finesse point are described in Saari (2005), so let me just state some conclusions. The first ensures that this point always exists, the second shows that the adjustment region $\mathbb{S}_d(q)$ can refine the standard concept of a "yolk," while the third asserts that the finesse point is an extension of the core.

THEOREM 5 *(Saari, 2005) With n voters given by their ideal points, the following are true:*

1 *For any q-rule and any number of issues, a q-rule d-finesse point exists.*

2 *For n voters and $n - 1$ issues, the adjustment set $\mathbb{S}_d(n - 1)$ is a proper subset of the yolk. In particular, for three voters and the majority vote, the adjustment region always is in the interior of the yolk.*

3 *A d-finesse point is a core point if and only if $d = 0$.*

7.   Selective Core and Finesse Point

The core and finesse point respond to *all possible* winning coalitions. But a quick glance at contemporary politics reveals certain coalitions that never can occur. Some examples, involving extreme conservatives and liberals, can be so preposterous that it is not worth listing them. Yet the core and the d-finesse point include possible reactions to the impossible. The response is obvious.

DEFINITION 6 *For a specified list of winning coalitions $\mathcal{C}$, which does not include all winning coalitions, a* selective core point *is one that cannot be beaten by any coalition in $\mathcal{C}$. Similarly a* selective d-*finesse point is one that is defined by the winning coalitions in $\mathcal{C}$.*

To illustrate these terms, notice that the Fig. 4a ideal points require $\mathbb{C}(2, 3) = \emptyset$. While the core is empty, suppose that the issue represented by the horizontal axis is sufficiently divisive (maybe extreme views on abortion) to prevent a $\{1, 2\}$ coalition from forming. Thus $\mathcal{C} = \{\{1, 3\}, \{2, 3\}\}$, and the selective core consists of voter 3's ideal point.

8.   Conclusion

As established in Saari and Sieberg (2001), Saari (2006), and elsewhere, pairwise voting loses so much information that it should not be used. Included among the dismissed information is the rationality of voters as captured by their transitive ranking of preferences. The problem of using pairwise methods is captured by Arrow's impossibility theorem (Arrow, 1952); the resolution, which is achieved by reintroducing the rationality of voter preferences, is described in Saari (2001).

Beyond standard voting are game theoretic concepts such as the core. The core, however, fails to exist in many natural settings. We know this; if the core always did exist, we would be stymied forever at the status quo. On the other hand, we must understand when the core does exist and persist, and what to do when it does not. Beyond answering these questions, it is suggested why the prevalence of stability is greater than usually indicated by theory. On the other hand, there are settings, as captured by the chaos theorem, where stability does not exist. Here a sense of stability is reintroduced via the finesse point. This point is not a gift; it requires effort. Namely, the finesse point ensures a "minimal" (which could be "large") effort to counter new proposals. The relationship of this point to positive and negative campaigning is indicated in Saari (2005).

References

Arrow, K.J. [1952] 1963. *Social Choice and Individual Values* 2nd. ed., Wiley, New York.

Banks, Jeffrey S., "Singularity Theory and Core Existence in the Spatial Models." *Journal of Mathematical Economics* 24 (1995): 523–26.

Li, Linfang, and D. G. Saari, "Sen's Theorem: Geometric Proof and New Interpretations," IMBS Discussion papers, University of California, Irvine, 2005.

McKelvey, Richard, \General Conditions for Global Intransitivities in Formal Voting Models." *Econometrica* 47 (1979): 1085-1112.

McKelvey, Richard. "Structural Instability of the Core." *Journal of Mathematical Economics* 15 (1986): 179-98.

McKelvey, Richard and Norman Schofield. "Generalized Symmetry Conditions at a Core Point." *Econometrica* 55 (1987): 923-934.

Nurmi, Hannu. *Voting Procedures under Uncertainty*. New York: Springer, 2002.

Plott, Charles, "A Notion of Equilibrium and its Possibility Under Majority Rule." *American Economic Review* 57 (1967): 787-806.

Saari, D. G., *Basic Geometry of Voting*, Springer-Verlag, New York, 1995.

Saari, D. G., "Connecting and resolving Sen's and Arrow's Theorems," *Social Choice & Welfare* **15** (1998), 239-261.

Saari, D. G., "The generic existence of a core for q-rules." *Economic Theory* **9** (1997), 219-260.

Saari, D. G. *Decisions and Elections: Explaining the Unexpected*. Cambridge University Press, New York, 2001.

Saari, D. G., "Geometry of stable and chaotic discussion," *Amer. Math. Monthly,* **111** (May 2004), 377-393.

Saari, D. G. *Condorcet Lectures,* Université de Caen, November, 2004.

Saari, D. G., "Which is better: the Condorcet or Borda winner?," *Social Choice & Welfare*, 2006.

Saari, D. G., " 'Finessing' a point to augment the core," UCI preprint, October, 2005.

Saari, D. G., and A. Petron, Negative Externalities and Sen's Liberalism Theorem, *Economic Theory* **28** (June, 2006), 265-281.

Saari, D. G., and K. Sieberg, The sum of the parts can violate the whole, *Amer. Pol. Science Review* **95**, (Number 2, June 2001), 415-433.

Schofield, Norman, "General Instability of Majority Rule." *Review of Economic Studies* **50** (1983): 695-705.

Tataru, Maria, "Growth Rates in Multidimensional Spatial Voting". *Ph.D. Dissertation, Northwestern University,* 1996.

Tataru, Maria, "Growth Rates in Multidimensional Spatial Voting". *Mathematical Social Sciences* 37 (January, 1999): 253-63.

A Comparison of Electoral Formulae for the Faroese Parliament

Petur Zachariassen[1], Martin Zachariassen[2]

[1]University of the Faroe Islands

[2]Department of Computer Science, University of Copenhagen

Abstract   The Faroese electoral system uses a method of proportional representation for distributing the seats in the Faroese Parliament (The Løgting). The electoral formulae attempt to give each political party a number of seats that is close to its vote share. In addition, each district should receive a number of delegates that is proportional to the number of voters in the district. We show that the current electoral formula has significant weaknesses, and propose 7 alternative electoral formulae which consider various subsets of constraints – such as lower bounds on the number of seats in districts and electoral thresholds. Numerical simulations with the current and proposed electoral formulae on the elections from 1978 to 2004, and on randomly generated election results, are presented. The results show that some of the proposed alternatives clearly are superior to the current electoral formula with respect to well-known quality measures.

Keywords: Electoral formula; apportionment method; proportional representation; multi-member districts; biproportional rounding; numerical simulations.

1.   Introduction

The Løgting is the parliament of the Faroe Islands, a group of islands in the North Atlantic Ocean between Scotland, Norway and Iceland. The islands, with its nearly 50 thousands inhabitants, have been an autonomous region of the Kingdom of Denmark since 1948 and have, over the years, taken control of most matters, except amongst others defence and foreign affairs.

The Løgting is elected for a period of four years, and has 27 district seats plus up to 5 adjustment seats. Election can take place before the end of an election period if the Løgting agrees on dissolving itself, as was the case in the latest election in January 2004, where the previous election was held in April 2002.

For more than fifty years the members of the Løgting typically have belonged to one of 6 political parties, four large parties (6-8 seats each) and two smaller parties (1-3 seats each). Traditionally the Faroese government has been based on a parliamentary majority consisting of 2-3 large parties and sometimes one or both of the smaller parties. At the 2004 election, for instance, the party distribution in the Løgting was 8+7+7+7+2+1, and 3 of the larger parties with 7+7+7 seats formed the current coalition.

The electoral system of the Løgting can be classified as a List PR system with two-tier districting. At the lower-level 27 members are elected in 7 multi-member districts using the d'Hondt (Jefferson) apportionment method. District codes and district magnitude are: NO/4, EY/5, NS/2, SS/8, VA/2, SA/2, and SU/4. Then, at the national level, up to 5 adjustment seats are distributed among parties and districts based on the LR Hare (Hamilton) method (see Zachariassen (2005) for details). Note that with this two-tier method, the total number of seats can vary from 27 to 32. The electoral threshold is 1/27 for receiving an adjustment seat, while district seats are distributed with the threshold inherent to the d'Hondt method.

As part of the electoral formula that assigns the seat distribution of the Løgting a 'necessary' number of adjustment seats is calculated in order to obtain a 'perfect' proportional result. When this number happens to be less than 5, the size of the Løgting would be less than 32. However, at the nine general elections held since 1978, when the current Election Act came into force, the necessary number of adjustment seats turned out to be at least 6 (tree times) and at most 10 (twice), so the size of the Løgting topped at 32 at every election. Generally, the higher the number of necessary adjustment seats, the less proportional is the result. Furthermore, an over-representation at the district-level in the smallest district contributes to the high necessary number of adjustment seats.

This weakness of the electoral formula for the Løgting became obvious at the 2004 election, when the third largest party in vote share obtained 8 seats, one more than the two largest parties in vote share, obtaining 7 seats each. Such a vote-seat reversal is not suprising when taking into account the 10 adjustment seats that would be necessary in this election to get a proportional result. Consequently there seems to be a growing interest among politicians to revise the current Electoral Act.

The main purpose of this paper is to compare the current electoral formula with some alternatives while preserving the concept of multi-member districts. First we present a numerical comparison of some classical apportionment methods for proportional representation using a 32 seat parliament election with 6 running parties in a single national district (section 2). Then 8 district-based electoral formulae are presented, considering various subsets of constraints – such as lower bounds on the number of seats in the districts and electoral thresholds (section 3). Numerical simulations on the elections from 1978 to 2004, and on randomly generated election results, are presented in section 4. The generated elections results are based on the 2004 election and on a 10-year demographic forecast. Concluding remarks are given in section 5.

2. Seat Bias of Apportionment Methods

In this section we compare some well-known apportionment methods by simulating a 32 seat parliament election with 6 running parties in a single national district. No electoral threshold is used. The simulation procedure is as follows: A 5 pseudo-random sample $\{p_i, i = 1, ...5\}$ from the uniform distribution on $[0, 1]$ is generated, and, assuming $p_1 < p_2 < p_3 < p_4 < p_5$, the 6 parties vote shares are defined by the numbers $p_1, p_2 - p_1, p_3 - p_2, p_4 - p_3, p_5 - p_4$ og $1 - p_5$. The results given below are based on a series of 5000 elections simulated in this way.

Four apportionment methods are used on these simulated vote shares: d'Hondt (Jefferson), Scandinavian 'Modified Sainte-Laguë', Sainte-Laguë (Webster) and LR-Hare (Hamilton). For an introduction to these methods, see Balinski and Young (2001).

Figure 1 displays, for each method, the mean seat biases for each party. The parties are ordered from the smallest to the largest by their vote shares. Figure 1 also displays (for each method and party) the maximum seat bias of the actual seat share above and below the fair seat share according to the vote share. For example, by using the d'Hondt method for the next largest party the lowest occurrence among the 5000 simulated elections is approximately -0.8 (e.g. 7 seats with 7.8 in fair seat share).

Table 1 shows the mean bias values in Figure 1, along with – for the three classical methods d'Hondt, Sainte-Laguë, LR-Hare – conjectured formula values from Schuster et al (2003). There seems to be a fairly good correspondence between simulated and formula values. Simulated values for the Modified Sainte-Laguë are also shown.

For each election simulation we define the 'allocation distance' between two apportionment methods to be half the L_1-distance between their seat allocation vectors. Thus, when two methods result in identical seat allocations the distance is 0. The distance between the seat allocations (2,2,4,5,9,10) and (1,1,4,5,9,12), based on the same vote share, is 2.

Table 2 presents some statistics on the allocation distance between each pairwise combination (6 in total) of the four considered apportionment methods based on the 5000 simulated elections. A simple frequency distribution (f=0, 1, >1) is shown along with the empirical mean of the allocation distance. Figure 2 depicts an 'apportionment method map', where the pairwise geometrical distance approximately equals the empirical mean of the allocation distance shown in Table 2.

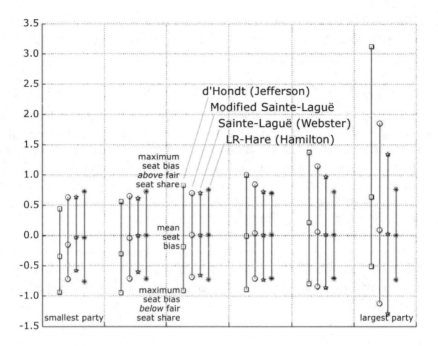

Fig. 1. Mean seat bias and maximum seat bias above and below the fair seat share for the apportionment methods: d'Hondt (Jefferson), Modified Sainte-Laguë, Sainte-Laguë (Webster), LR-Hare (Hamilton) based on 5000 simulations of a 32 seat parliament election with 6 running parties. The parties are ordered from the smallest (left) to the largest (right) by their vote shares. For example, by using the d'Hondt method the next largest party's lowest occurrence is approximately -0.8 (e.g. 7 seats with 7.8 in fair seat share).

Table 1. Mean seat bias values

Party rank	k	Mean fair seat share	d'Hondt		Modified Sainte-Laguë	Sainte-Laguë		LR-Hare	
			Simu-lations	For-mula*	Simu-lations	Simu-lations	For-mula	Simu-lations	For-mula
largest	1	12.36	0.636	0.725	0.095	0.022	0.021	0.005	0.009
	2	7.75	0.214	0.225	0.057	0.010	0.011	0.009	0.009
	3	5.29	-0.009	-0.025	0.038	0.007	0.006	0.012	0.009
	4	3.50	-0.187	-0.192	0.010	0.002	0.003	0.011	0.009
	5	2.13	-0.306	-0.317	-0.045	-0.001	-0.000	-0.002	0.009
smallest	6	0.97	-0.347	-0.417	-0.155	0.039	-0.041	-0.035	-0.045

$$^*(-1 + \sum_{j=k}^{f} \tfrac{1}{j})/2, \ f = 6 \ \text{(Schuster et al, 2003)}$$

Table 2. Allocation distance between each pairwise combination (6 in total) of the four considered apportionment methods based on 5000 simulations of a 32 seat parliament election with 6 running parties. The combinations are numbered from the largest to the smallest distance. Frequency distributions (f=0, 1, >1) are shown, e.g. in 16% of the simulations Sainte-Laguë and Modified Sainte-Laguë produce allocations with distance 1.

	Pair of methods	Mean distance	f=0 %	f=1 %	f>1 %
1	d'H – LR-H	0.935	24	59	17
2	d'H – S-L	0.924	27	55	18
3	d'H – M-S-L	0.791	33	55	12
4	LR-H – M-S-L	0.281	73	26	1
5	S-L – M-S-L	0.172	83	16	1
6	S-L – M-S-L	0.168	83	17	0

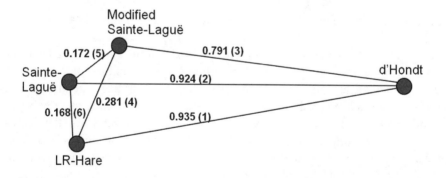

Fig. 2. Apportionment method map based on the allocation distances in Table 2. The largest distance is between d'Hondt and LR-Hare, the shortest between Sainte-Laguë and LR-Hare.

The apportionment method map in Figure 2 confirms the well-known similarity of LR-Hare and Sainte-Laguë; in 83% of the simulations these two apportionment methods produce identical allocations (Table 2). The outlying of d'Hondt is also to be expected; in only 24% and 27% of the simulations this apportionment method produces the same allocations as respectively LR-Hare and Sainte-Laguë.

In summary, the simulations based on a scenario with a 32 seat parliament with 6 running parties confirm the differences and similarities of four well-known apportionment methods (Figure 2). To some extent the simulations also quantifies the implied seat bias compared to the fair seat share by using these methods (Figure 1). The mean seat bias of d'Hondt method varies from -0.35 (for the smallest party) to +0.64 (for the largest party). Also one observes that the mean seat bias for Sainte-Laguë and LR-Hare are neglible for all the 6 parties in the simulations, but there are differences in the spread for the two methods; Sainte-Laguë shows a larger spread than LR-Hare for the two largest parties – for the three smallest parties the relation is reversed.

3.   District-Based Electoral Formulae

In this section we present a number of different district-based electoral formulae for distributing the seats in the Faroese Parliament. The formulae are compared by performing numerical simulations on the elections from 1978 to 2004, and on randomly generated election results. In section 3.1 the method for constructing randomly generated election results is described.

The electoral formulae are compared by computing three performance indices, which measure the 'error' of the seat distributions when compared to the underlying vote distributions. The indices are described in detail in section 3.2.

In section 3.3 we present 8 different electoral formulae, including the electoral formulae that is used today. Controlled rounding and the biproportional divisor method are among the other investigated formulae. In their purest form, these two methods have no supplementary constraints such as electoral thresholds or lower bounds on the number of district seats; they compute seat distributions which are optimal within their framework. Therefore, the results achieved by these methods are basically the best possible when (bi)proportionality is the only objective.

The results of our numerical simulations with the 8 electoral formulae are discussed in section 4. The results are primarily discussed by comparing performance indices. Detailed seat distributions for each of the elections 1978-2004 are available from www.nvd.fo/fileadmin/pdf/Appendix.pdf.

3.1   Vote Distributions

Three sets of vote distributions have been studied. The first set contains the 9 election results from 1978 to 2004. These are the elections where the current electoral formulae has been used. Numerical simulations on these vote distributions show what the outcome of an election would have been (wrt. seats) if an alternative electoral formulae had been used.

A more thorough comparison of the electoral formulae has been achieved by performing simulations on randomly generated election results. These election results were obtained as follows: Let a_{ij} be the number of votes for party i in district j in the 2004 election. In the randomly generated election result, the number of votes given to party i in district j was first chosen uniformly at random from the interval $[0.5 * a_{ij}; 1.5 * a_{ij}]$. Then for each district the generated vote numbers were scaled such that the total number of votes in the district became identical to the total number in the 2004 election. Thus we chose to fix the number of voters in each district, but to vary the number of votes given to each party. We generated 100 random election results of this type, and they formed the second set of vote distributions.

In order to predict how the electoral formulae will perform in the future we generated a third set of vote distributions based on an estimate on the number of voters in each district in 2014. Otherwise the same method as for the randomly generated 2004 election results was used. The expected population changes for each district were -3% (NO), +2% (EY), +1% (NS), +6% (SS), –3% (VA), –8% (SA) and –7% (SU).

3.2   Performance Indices

Since no single performance index gives the full picture of the performance of an election formulae, we have chosen to report the values of three different indices.

Consider an apportionment for n parties, where party i with quota q_i has received t_i seats, $i = 1, \ldots, n$. Quota and seat numbers are given in percentages:

$$\sum_i q_i = \sum_i t_i = 100$$

Gallagher/least-squares index.   This index is basically the Euclidean distance between the quota and seat vectors, that is, the square root of the sum of squares of the differences between quota and seats (Gallagher, 1991):

$$\sqrt{\frac{1}{2} \sum_i (q_i - t_i)^2}$$

A small Gallagher index value indicates that the seat numbers are close to the quota numbers. The Gallagher index is similar to the Loosemore-Hanby index (Loosemore and Hanby, 1971), which measures the sum of absolute differences between quota and seat numbers. The Gallagher index is, however, more sensitive to large deviations between seat and quota numbers. A third index is the Chi-square index, which is the relative sum of squares. It can be shown to measure the extent to which the voters are equally represented (Pukelsheim, 2000).

The Gallagher index and the other two indices are strongly correlated. Since the Gallagher index also is the most widespread, we have chosen to focus on the Gallagher index in this study.

Quota breaches.   A breach of quota happens when the difference between quota and seats is 1 or more. The quota breach index counts the number of quota breaches for all parties. Note that this index is always 0 for quota based methods such as the LR-Hare (Hamilton) method or controlled rounding.

Bias.   Some methods are 'biased', which means that they give small or large parties an advantage. In the Faroese Parliament case the number of parties is small, so well-known methods for measuring bias are not suitable. The following index was therefore chosen for measuring bias. Let i_1 and i_2 be the indices for the two parties that have received the largest number of votes. Our bias index is now computed as the total number of seats for these two parties divided by their total quota:

$$\frac{t_{i_1} + t_{i_2}}{q_{i_1} + q_{i_2}}$$

If this number is greater than 1, then the large parties have an advantage over the smaller parties, and vice versa if the index is less than 1.

In the discussion above the indices were computed for parties, but we may compute similar indices for districts. Here q_i and t_i are the quota and seat numbers, respectively, for district i. In order to distinguish between indices based on parties and districts, we use the notion party-indices and district-indices in the discussion of the results in section 4.

3.3   Electoral Formulae

Our study covers 8 electoral formulae, ranging from the (constrained) current electoral formulae to unconstrained biproportional divisor methods and controlled rounding. Each formula (or method) has a number which is given below and which is used for identification in tables and figures.

Current electoral formula (method 1).   The two-tier method given by the law from 1978. A total of 27 district seats (4+5+2+8+2+2+4) are first distributed independently within each district using the d'Hondt divisor method (lower-tier method). Then up to 5 adjustment seats are distributed among parties and districts (upper-tier method); the procedure for distributing adjustment seats is based on the largest-remainder (Hamilton) method (see Zachariassen (2005) for details). Note that with this two-tier method, the total number of seats can vary from 27 to 32 seats. The electoral threshold is 1/27 for the distribution of adjustment seats, while district seats are distributed without any threshold.

Current electoral formula with Sainte-Laguë (method 2).   Similar to method 1, but the 27 district seats are distributed using the Sainte-Laguë divisor method.

Hylland method with and without threshold (methods 3 and 5).   As in method 2, the 27 district seats are distributed using the Sainte-Laguë divisor method. The distribution of adjustment seats follows a method suggested by Hylland (1990); as a minor change, we fixed the number of adjustment seats to exactly 5 (as compared to at most 5), such that the total number of seats always becomes 32. Two variants were considered: One that has an electoral threshold when distributing adjustment seats (method 3), and one that has no threshold (method 5).

The Hylland method for distributing adjustment seats is as follows: First the total number of district seats is computed for each party. Then the number of adjustment seats for each party is computed by using the socalled adjusting Sainte-Laguë divisor method. This means that Sainte-Laguë is 'warm-started' with the number of district seats for each party as input. When given enough adjustment seats, this method will end up with a true Sainte-Laguë distribution, but since we (only) distribute 5 seats, the result may not be a true Sainte-Laguë distribution.

When the number of adjustment seats for each party has been found, these seats are then distributed among the districts independently for each party. Again the adjusting Sainte-Laguë method with the district seat distribution as input is used for each party.

The Hylland method for distributing adjustment seats is significantly simpler than the one currently used, which mixes several different paradigms.

Balinski with/without constraints and threshold (methods 4, 6 and 7).
The biproportional divisor method given by Balinski and Demange (1989a/b), Balinski and Rachev (1997), and Balinski (2002). Sainte-Laguë (or standard) rounding was employed. Algorithmic and experimental aspects of divisor-based biproportional rounding methods are discussed in Maier (2006) and Zachariasen (2006). The biproportional divisor method – hereafter just called

the Balinski method – does not distinguish between district and adjustment seats.

The Balinski method was used for the first time ever at the Zürich City Parliament election on February 12, 2006. The application of the biproportional divisor method in Zürich is outlined in Pukelsheim and Schuhmacher (2004).

We evaluated three variants of the Balinski method. In the unconstrained variant (method 7), we distributed the marginals using Sainte-Laguë according to row and column vote sums; no electoral threshold was used. In the constrained variant, we distributed the district marginals by using the current distribution of district seats (4+5+2+8+2+2+4) as input to the adjusting Sainte-Laguë method. This way each district receives at least as many seats as it does using the current electoral formula. We considered one constrained variant with an electoral (party) threshold of 1/27 (method 4), and one constrained variant without a threshold (method 6).

Controlled rounding (method 8).   The controlled rounding algorithm (Cox and Ernst, 1982) is a quota method, and it is the two-dimensional generalization of the one-dimensional LR-Hare (Hamilton) method. Basically the method minimizes the total rounding error for all elements in the quota matrix, including the marginals. We only considered the purest variant of this algorithm, i.e., without electoral threshold or other constraints.

4.   Results

First we present some results for the 1978-2004 election data. Figure 3 (left) shows the Gallagher party-index for each of the 8 electoral formulae. For the current electoral formula (1), the Gallagher party-index is in the range 2 to 4 – with a slight increase from 1978 to 2004. For the other methods that have an electoral threshold (2,3,4), the Gallagher party-index is a bit smaller. The constrained Balinski method (4) has the smallest Gallagher party-index among the methods with electoral thresholds. Among the remaining methods (5,6,7,8), the Balinski methods (6,7) and controlled rounding are more or less equal with a Gallagher party-index of approximately 2.

The Gallagher district-indices are presented in Figure 3 (right). The current and Sainte-Laguë based electoral formulae (1,2) and the Hylland methods (3,5) have significantly larger Gallagher district-indices than the other methods. The reason is that these methods distinguish between district and adjustment seats. The constrained Balinski methods (4,6) have the same lower bound on the number of seats in each district as the current electoral formula, but there is no distinction between district and adjustment seats. There is a general increase in the Gallagher district-indices from 1978 to 2004. This is a result of an increased imbalance in the 'value' of district seats from 1978 to 2004. Apparently the adjustment seats cannot correct this imbalance.

The number of quota breaches presented in Figure 4 has the same tendency as the Gallagher index. For the 9 elections from 1978 to 2004, the current electoral formula had 3 quota breaches for the parties. The Balinski methods (4,6,7) had no quota breaches. For the bias index, the current formula also gives the worst result for parties – mainly because of the use of d'Hondt when distributing district seats.

Table 3 and Figures 5/6 give the results for the generated 2004 election data. In the figures the distribution of the indices over the 100 generated election results is shown. The Gallagher indices for the 8 electoral formulae are presented in Table 3 (top) and Figure 5. The current electoral formula (1) has an average Gallagher party-index of 3.31, which is higher than for all other methods. The electoral threshold has a significant influence on the Gallagher party-index; the four methods that do have an electoral threshold (1,2,3,4) clearly have higher indices than the remaining methods which do not have an electoral threshold. For the Gallagher district-indices we see a similar pattern. Those methods that have lower bounds on the number of district seats perform significantly worse with respect to the Gallagher district-index. Note that the district-index distributions are less Gaussian, since we fixed the number of votes for each district in our generation procedure.

The results on quota breaches for the generated 2004 elections are given in Table 3 (middle) and Figure 6. Again the current electoral formula (1) has more quota breaches, both for parties and districts, than the other methods. The Balinski methods (4,6,7) have very few quota breaches, since their marginal distributions are based on Sainte-Laguë.

The results for the generated 2014 election data are similar to the 2004 results, but there is one interesting difference. For the methods that have lower bounds on the number of district seats, the Gallagher district-indices are much higher for 2014 than for 2004. The reason is the increased imbalance in the 'value' of district seats.

5.   Conclusion

In this paper we presented substantial numerical simulations with proportional and biproportional apportionment methods for the Faroese Parliament. Our simulations with classical apportionment methods for a 32 seat parliament confirmed that the Sainte-Laguë and LR-Lare methods have the smallest seat biases (when compared to the fair seat share). For the district-based, or biproportional problem, we made simulations with 8 different methods. When compared to the current electoral formula for the Faroese Parliament, most of the alternatives were superior with respect to well-known quality measures, such as the Gallagher index. Our study supports the public sentiment to replace the current electoral formula with a divisor-based biproportional method.

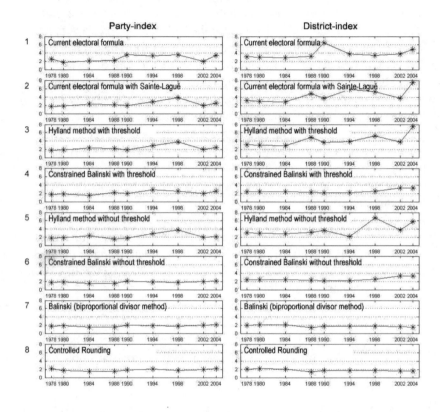

Fig. 3. Graphs for the **Gallagher index** for the nine Faroese Parliament elections from 1978 to 2004. 8 different electoral formulae are used. Party-index to the left and district-index to the right.

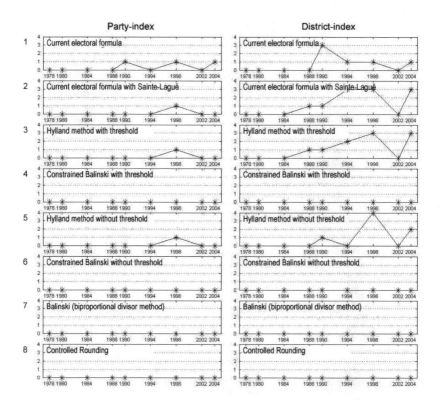

Fig. 4. Graphs for the **quota breach index** for the nine Faroese Parliament elections from 1978 to 2004. 8 different electoral formulae are used. Party-index to the left and district-index to the right.

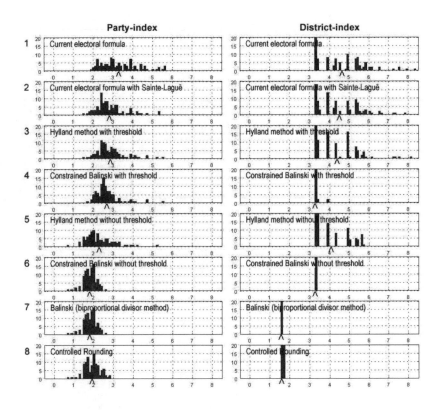

Fig. 5. Histograms for the **Gallagher index** for 100 simulations based on the Faroese Parliament election 2004. 8 different electoral formulae are used. Party-index to the left and district-index to the right. The ^ sign marks the empirical mean value.

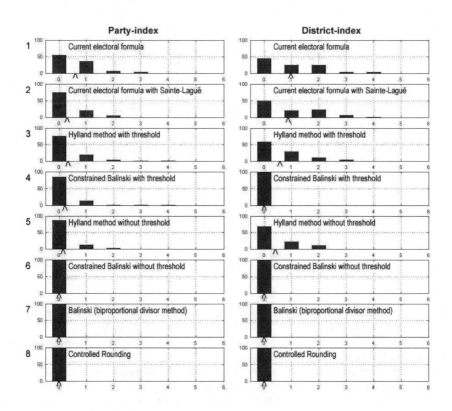

Fig. 6. Histograms for the **quota breach index** for 100 simulations based on the Faroese Parliament election 2004. 8 different electoral formulae are used. Party-index to the left and district-index to the right. The ^ sign marks the empirical mean value.

Table 3. Performance indices for 100 simulated vote distributions based on the 2004 election for the Faroese parliament.

Index		Electoral Formula	Party-index				District-index			
			mean	1)	lowest	highest	mean	1)	lowest	highest
Gallagher	1	Current electoral formula	3.31	-	1.90	5.60	4.64	-	3.27	8.26
	2	Current electoral formula with S-Laguë	2.83	.	1.46	5.35	4.52	-	3.27	7.75
	3	Hylland method with threshold	2.86	.	1.46	5.50	4.38	-	3.27	8.07
	4	Constrained Balinski with threshold	2.68	.	1.73	5.50	3.29	-	3.27	3.86
	5	Hylland method without threshold	2.30	.	0.73	5.23	4.09	-	3.27	5.73
	6	Constrained Balinski without threshold	1.82	+	0.73	2.64	3.28	-	3.27	3.28
	7	Balinski (biproportional divisor method)	1.82	+	0.73	2.64	1.57	+	1.57	1.58
	8	Controlled Rounding	1.92	+	0.73	2.83	1.60	+	1.57	1.66
Quota	1	Current electoral formula	0.60	-	0	3	0.98	-	0	4
breaches	2	Current electoral formula with S-Laguë	0.30	.	0	2	0.88	-	0	4
	3	Hylland method with threshold	0.34	.	0	4	0.58	-	0	3
	4	Constrained Balinski with threshold	0.22	.	0	4	0	+	0	0
	5	Hylland method without threshold	0.16	.	0	2	0.42	.	0	2
	6	Constrained Balinski without threshold	0	+	0	0	0	+	0	0
	7	Balinski (biproportional divisor method)	0	+	0	0	0	+	0	0
	8	Controlled Rounding	0	+	0	0	0	+	0	0
Bias	1	Current electoral formula	1.03	.	0.89	1.13	0.92	-	0.85	0.96
	2	Current electoral formula with S-Laguë	1.02	+	0.93	1.12	0.92	-	0.85	0.96
	3	Hylland method with threshold	1.02	+	0.94	1.11	0.92	-	0.80	0.96
	4	Constrained Balinski with threshold	1.02	+	0.97	1.11	0.96	.	0.90	0.96
	5	Hylland method without threshold	0.99	+	0.89	1.08	0.93	.	0.85	0.96
	6	Constrained Balinski without threshold	0.99	+	0.93	1.04	0.96	.	0.96	0.96
	7	Balinski (biproportional divisor method)	0.99	+	0.92	1.04	1.01	+	1.01	1.01
	8	Controlled Rounding	1.00	+	0.93	1.08	1.01	+	1.01	1.01

1) Descriptions

	Gallagher (G)		Quota breaches (Q)		Bias (B)
+	G < 2.00	+	Q = 0	+	0.98 < B < 1.02
.	2.00 < G < 3.00	.	0 < Q < 0.50	.	0.95 < B < 0.98 or 1.02 < B < 1.05
-	G > 3.00	-	Q > 0.50	-	B < 0.95 or B > 1.05

References

Balinski, M. L., 2002. Wahlen in Mexico – Verhältniswahlrecht häappchenweise, *Spektrum der Wissenschaft*, Oktober, 72-74.

Balinski, M. L., and Demange, G., 1989a. An axiomatic approach to proportionality between matrices. *Mathematics of Operations Research* 14, 700-719.

Balinski, M. L., and Demange, G., 1989b. Algorithms for proportional matrices in reals and integers. *Mathematical Programming* 45, 193-210.

Balinski, M. L., and Rachev, S. T., 1997 Rounding proportions: Methods of rounding. *Mathematical Scientist* 22, 1-26.

Balinski, M. L., and Young, H. P., 2001. *Fair representation: Meeting the ideal of one man, one vote.* Brookings Institution Press, 2nd edition.

Cox, L. H., and Ernst, L. R., 1982. Controlled rounding. *INFOR* 20 (4), 423-432.

Gallagher, M., 1991. Proportionality, disproportionality, and electoral systems. *Electoral Studies* 10, 33-51.

Hylland, A., 1990. Lagtinget pa Færøerne: Beskrivelse av og kommentarer til valgordningen. Handelshøyskolen BI.

Loosemore, J., and Hanby, V. J., 1971. The theoretical limits of maximal distortion: some analytic expressions for electoral systems. *British Journal of Political Science* 1:4, 467-477.

Maier, S., 2006. Algorithms for Biproportional Apportionment Methods. In *Mathematics and Democracy. Recent Advances in Voting Systems and Collective Choice*, In Press, New York.

Pukelsheim, F., 2000. Mandatszuteilungen bei Verhäaltniswahlen: Erfolgswertgleichheit der Wählerstimmen. *Allgemeines Statistiches Archiv* 84, 447-459.

Pukelsheim F., and Schuhmacher, C., 2004. Das neue Zürcher Zuteilungsverfahren für Parlamentswahlen. Aktuelle Juristische Praxis – *Practique Juridique Actuelle* 5/2004, 505-522.

Schuster, K., Pukelsheim, F., Drton, M., and Draper, N. R., 2003. Seat biases of apportionment methods for proportional representation. *Electoral Studies* 22, 651-671.

Zachariasen, M., 2006. Algorithmic Aspects of Divisor-Based Biproportional Rounding. Technical Report 06-05, Department of Computer Science, University of Copenhagen.

Zachariassen, P., 2005. Uppgerð av løgtingsvali – eitt sindur um bygnað og virknað [in Faroese]. *Ársfrágreiðing* 2004, Fróðskaparsetur Føroya.

Zachariassen, P., and Zachariassen, M., 2005. A comparison of electoral formulae for the Faroese Parliament (The Løgting) [in Faroese]. Technical report *NVD-rit 2005:1, Náttúruvísindadeildin, Fróðskaparsetur Føroya*.

List of Talks

Invited Talks

M. Balinski: Fair majority voting (or: How to eliminate gerrymandering)

P. Edelmann: Measuring representation and apportionment in the United States

F. Aleskerov: Power indices using agents' preferences: theory and applications

M. Kilgour: A minimax procedure for electing committees

B. Simeone: The Sunfish against the Octopus: opposing compactness to gerrymandering

F. Pukelsheim: Current political issues of apportionment methods: preserving majority, securing minimum representation, obtaining double proportionality

D. Chaum: Multiparty computations characterizing cryptographic protocols

E. Giovannini: Evidence-based collective decision making: the role of Statistics

S. Brams: Critical strategies under approval voting: what gets ruled in and ruled out

H. Nurmi: Distance from consensus: a theme with variations

M. I. Shamos: The ten top problems in practical electronic voting

D. Chaum: Punchscan illustrating the power of crypto in voting

V. Ramírez: Proportional social choice

M. Salles: Further results on the stability set of voting games

Contributed Talks

A. Palomares: Thresholds of the divisor methods

S. Maier: Algorithms for biproportional apportionment methods

A. Pennisi: A wrong biproportional apportionment procedure: the case of Italy

P. Zachariassen: A comparison of electoral formulae for the Faroese Parliament

V. Fragnelli: A Simulative Approach for Evaluating Electoral Systems

C. Klamler: Choice Functions, Binary Relations and Distances

R. Sanver: Social choice when approval is an intrinsic part of individual preferences

List of Participants

Invited Speakers

Fouad Aleskerov, Russian Academy of Sciences, Moscow, Russia
Michel Balinski, École Polytechnique, Paris, France
Steven Brams, New York University, New York, USA
David Chaum, Katholik Universiteit, Louvain, Belgium
Paul Edelman, Vanderbilt University, Nashville, USA
Enrico Giovannini, OECD, Paris, France
Marc Kilgour, Wilfrid Laurier University, Waterloo, Canada
Jack H. Nagel, University of Pennsylvania, Philadelphia, USA
Hannu Nurmi, University of Turku, Finland
Friedrich Pukelsheim, Universität Augsburg, Germany
Victoriano Ramírez González, Universidad de Granada, Spain
Donald Saari, University of California, Irvine, USA
Maurice Salles, Universitée de Caen, France
Michael Ian Shamos, Carnegie-Mellon University, Pittsburgh, USA
Bruno Simeone, Università La Sapienza, Rome, Italy

Further Participants

Lorenzo Cioni, University of Pisa, Italy
Blanca Luisa Delgado Márquez, Universidad de Granada, Spain
Christian Klamler, Technische Universität, Graz, Austria
Isabella Lari, Università La Sapienza, Rome, Italy
Sebastian Maier, Universität Augsburg, Germany
Maria Luisa Márquez García, Universidad de Granada, Spain
Antonio Palomares Bautista, Universidad de Granada, Spain
Aline Pennisi, electoral systems expert, Rome, Italy
Federica Ricca, Università La Sapienza, Rome, Italy
Remzi Sanver, Bilgi University, Istanbul
Andrea Scozzari, Università La Sapienza, Rome, Italy
Petur Zachariassen, University of the Faroe Islands

Scientific Co-Directors of the Workshop

B. Simeone, F. Pukelsheim

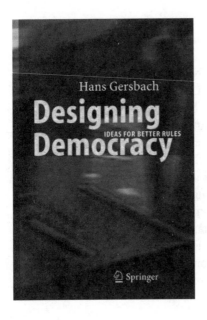

Hans Gersbach
Designing Democracy
Ideas for Better Rules
2005, XI, 243 p., Hardcover
ISBN: 3-540-22402-5

This book presents a number of ideas for drawing up new rules to improve the functioning of democracies. The first part examines ways of combining incentive contracts with democratic elections. Such a judicious combination can alleviate a wide range of political failures without impairing the principles on which democracies are founded. The second part presents new rules for decision-making, agendas and agenda settings which can transcend the limitations of prevailing democracies in achieving desirable outcomes. An example is flexible majority rules where the size of the majority depends on the proposal. The book comprises a sequence of simple models and intuitive explanations of the results they yield.

Printing: Krips bv, Meppel
Binding: Stürtz, Würzburg